# Biometric State

Biometric identification and registration systems are being proposed by governments and businesses across the world. Surprisingly they are under most rapid, and systematic, development in countries in Africa and Asia. In this ground-breaking book Keith Breckenridge traces how the origins of the systems being developed in places like India, Mexico, Nigeria and Ghana can be found in a century-long history of biometric government in South Africa, with the South African experience of centralised fingerprint identification unparalleled in its chronological depth and demographic scope. He shows how empire, and particularly the triangular relationship between India, the Witwatersrand and Britain, established the special South African obsession with biometric government, and shaped the international politics that developed around it for the length of the twentieth century. He also examines the political effects of biometric registration systems, revealing their consequences for the basic workings of the institutions of democracy and authoritarianism.

KEITH BRECKENRIDGE is an Associate Professor and the Deputy Director at the Wits Institute for Social and Economic Research at the University of the Witwatersrand.

# Biometric State

*The Global Politics of Identification and Surveillance in South Africa, 1850 to the Present*

Keith Breckenridge

*University of the Witwatersrand*

CAMBRIDGE
UNIVERSITY PRESS

# CAMBRIDGE
## UNIVERSITY PRESS

University Printing House, Cambridge CB2 8BS, United Kingdom

Cambridge University Press is part of the University of Cambridge.

It furthers the University's mission by disseminating knowledge in the pursuit of education, learning and research at the highest international levels of excellence.

www.cambridge.org
Information on this title: www.cambridge.org/9781107434899

First published 2014
First paperback edition 2016

*A catalogue record for this publication is available from the British Library*

*Library of Congress Cataloguing in Publication data*
Breckenridge, Keith, author.
Biometric state : the global politics of identification and surveillance in South Africa, 1850 to the present / Keith Breckenridge, University of the Witwatersrand.
    pages   cm
Includes bibliographical references and index.
ISBN 978-1-107-07784-3 (hardback)
1. Citizenship–South Africa–History.   2. Biometric identification–Government policy–South Africa–History.   3. Biometric identification–Political aspects–South Africa–History.   4. South Africa–Politics and government–19th century.   5. South Africa–Politics and government–20th century.   I. Title.
JQ1983.B74 2014
323.44830968–dc23
2014022526

ISBN 978-1-107-07784-3 Hardback
ISBN 978-1-107-43489-9 Paperback

For Catherine and Alexandra, who remind me that writing is not life

# Contents

# Figures

# Preface and acknowledgements

This book is a history of biometric government before computers; it argues that most, although not all, of the authoritarian political results of universal and centralised fingerprinting will persist into the epoch of computerised automation unless we pay special attention to them. It is also a new and probably a controversial interpretation of South African history. This is because the book sidesteps two large and established bodies of writing in its explanation of the peculiar course of South African history – the older liberal historiography on the politics of Afrikaner Nationalism and the newer (although now also old) Marxist studies of the ideological and institutional effects of mining-driven capitalism. I have chosen, instead, to follow the local effects of globally staged debates in the science and technology of biometrics in accounting for the Apartheid state, and its immediate aftermath. This is not because I reject the explanatory power and value of the liberal and Marxist interpretations of South African history. On the contrary, as I hope my other work shows, I am utterly persuaded of the tremendous analytical power of twentieth-century South African historical writing in its published and unpublished forms. My reasons for ignoring the well-formed pattern of the debate lie elsewhere.

My interest in this book has been in the global significance of South African history. Why does South African history matter? An easy answer is to say, as many do, that it does not, or not any more than any other mid-sized country. I do not agree, of course. And I want to look carefully at the claims being made by writers such as Arendt, Cell, Mazower, Mitchell and Woodward that South Africa marks the capstone of imperialism, of a system of racist bureaucracy that was applied in many other societies on the Atlantic basin. Biometric forms of identification, often formed by expansive international scientific debate and widely dispersed international experiments, lie at the heart of the story of South African history. And they are also instruments of global significance, especially in our time, for the transmission – around the former colonial world – of some of the distinctive features of the South African state. This is, mostly, not a good thing. I hope to entice the students of those other societies to consider the implications of South African history very carefully.

I also have in mind a more local scholarly project. Like the academic writing of many other societies, South African historiography has dug itself into something of an intellectual hole over the last three decades. Some of this is intellectual solipsism, nurtured by a tumultuous national politics that requires detailed familiarity with organisations, individuals and cultural politics that is, at best, difficult to translate. The publishing imperative that applies in South Africa, which effectively only rewards the publication of research articles, does little to encourage wide and deep comparative reading. And the turn, after Edward Thompson's attack on Althusser, to a fine-grained social history driven by the explanatory power of deep archives and very interesting oral sources has encouraged provincial and local research that has been another source of this isolation. But there is, also, an implicit and not often articulated view that South African history is *sui generis*, completely distinct from any other society, and disconnected by its history from both its neighbours and its distant colonial peers. This argument has contributed significantly to the growing global isolation of South African historical writing. My point here is to reverse this argument – that the peculiarity of our history is derived from its connections with the wider imperial world, and that those linkages provide the basis for very interesting and productive comparisons. These relationships also speak very usefully to our contemporary interest in the dim prospects for democracy and social justice in the former colonial world.

In my attempts to understand South Africa and the other societies shaped by the obsessions of biometric government I have indulged a taste for wide reading, sometimes, I suspect, to the concern of my colleagues. Seema Maharaj, the Interlibrary Loan Librarian at my former university, provided an efficient and, most importantly, free stream of exotic books. I owe considerable thanks to Jane Caplan and David Lyon for their guidance, and for exposure to the members of their respective Identinet and Surveillance Studies collaborations. I cannot name all the participants in the many workshops and seminars that have been assembled by these scholarly networks but they have enormously influenced my understanding of the global forms of identification and surveillance. I am also sure that they will continue to find abundant reason to correct me.

I would like to thank my colleagues at WISER, especially Belinda Bozzoli, Achille Mbembe and Sarah Nuttall, who, in very different ways, have worked to make Johannesburg a remarkable place from which to study the world. My thanks to Michael Watson and Cambridge University Press for finding a place for this book, and for their easy professionalism.

Much of the book was written in two periods of splendid isolation in Cambridge, for which I owe thanks to St John's College, for a Colenso Visiting Scholarship, and to Sujit Sivasundaram and Simon Schaffer. From Ann Arbor, I would like, especially, to thank my friends Gabrielle Hecht and Paul Edwards for ongoing institutional and scholarly support. My colleagues in Durban – Marijke du Toit, Vukile Khumalo, Thembisa Waetjen and Goolam Vahed – were persistently encouraging and demanding. I offer special thanks to all the members of the History and African Studies Seminar. To Suryakanthie Chetty, Bernard Dubbeld, Prinisha Badassy, Vanessa Noble, Paul Rouillard, Nafisa Essop Sheik and Stephen Sparks for gathering evidence of many kinds. And to Charles van Onselen for his help over many years. David William Cohen and Jeff Guy have been encouraging, perplexing and brave colleagues and collaborators. We will continue to debate, I hope, how history should be written and what it was actually about. And then, finally, my largest intellectual and institutional debts are owed to Simon Szreter. All of these people, of course, have no responsibility for the tendentious arguments and errors in the pages that follow.

This book is dedicated to Catherine Burns and Alexandra Breckenridge, with whom I have shared the pleasures and stresses of running a household filled with visiting academics, a very busy young woman, and three dogs. Book writing, under those circumstances, is a wicked indulgence.

# Introduction: the global biometric arena

Over the last century South Africa has acted as a global stage for the schemes and controversies of biometric government. This history has been tumultuous, but it has also, after four generations, produced a blueprint for a new architecture of state power that is spreading through the former colonies of the European empires – in these territories a clearly distinguishable biometric state is taking form, which, for the moment at least, seems to mark the technological apogee of the information society. Around the world biometric registration systems are changing the way a particular group of states undertake vital registration, build voters' rolls, distribute welfare benefits, control credit transactions, issue identity documents and police immigration. These new states, and their citizens, are adopting technologies and structures of administration that were often first developed and most fully elaborated in South Africa. They are also taking sides – often unwittingly – in an international political argument about the virtues and faults of biometric identification that has been animated, in every generation, by the twentieth-century history of South Africa.

This, then, is the history of a new kind of state, a biometric state. The explanation I offer is necessarily double-sided, Janus-faced, with lines of causality that run in opposing directions. On the one side, like other transnational histories, the book examines the ways in which the world made South Africa, in particular how the global fingerprinting project created a distinctive state in this country.[1] On the other, it examines how the events and ideologies produced by the very local (and often obscure, antipodean) struggles of this history around biometric identification fashioned a global politics.

In this introduction my goal is to orient my readers to the project of the book, and to clear – or at least identify – the most serious obstacles to my broader argument. The chapter falls into two obvious halves: the first

[1] E. Rauchway, *Blessed Among Nations: How the World Made America* (New York: Hill and Wang, 2007).

1

deals with problems of nomenclature and the second with the overall argument of the book. To begin with I ask, what is the state? And what, perhaps more obviously, are biometrics? The answers, inevitably, are not straightforward in either case but I try to venture working explanations for each of them. I then turn to the significance of what happens when the two terms are combined in the biometric state. This is a new kind of state, with a distinctive form of organisation, one which requires us to depart in important ways from the most influential ways of thinking about bureaucratic power. As an example of the biometric state I offer a very brief discussion of the current Aadhaar project in India, which has technological, philosophical and political links to the South African history. The chapter then turns to an overview of the story presented in the book.

### The problem of the state

As this is the story of the emergence of a new state, a working understanding of what I mean by that word seems a good place to begin. This is not, unfortunately, as straightforward as it might seem. The first problem has its origins in an old enthusiasm in European philosophy for granting personality to the state. In these accounts the state is *reified*, treated as a god-like person, freed from the rules of morality that apply to human individuals. We can track the power that follows from attributing autonomous moral interests to the state in the ongoing hold of Machiavelli's 500-year-old argument. The consequences of his recommendation of a new kind of mercilessly enforced political virtue, one that seeks only to defend the interests of the state – regardless of conventional morality – can be seen in our own time and in every generation before it; in this *raison d'état* the state becomes a person with an unchecked prerogative for self-defence.[2] This idea of an autonomous – god-like – personality was bolstered by Hobbes' *Leviathan*[3] and, especially, by Hegel's account of the state as the expression of a universal idea, carrying

---

[2] Niccolò Machiavelli, *The Prince*, vol. 36 (London: Oxford University Press, 1903), 41–3, 69–71; Quentin Skinner, *Machiavelli: A Very Short Introduction* (Oxford University Press, 2000), 34–81. On the hegemony, in Italy, of the *raison d'état* over older forms of personal political virtue, see Maurizio Viroli, *From Politics to Reason of State: The Acquisition and Transformation of the Language of Politics 1250–1600* (Cambridge University Press, 2005), 237–80. Machiavelli insisted, in *Discourses on the First Decade of Titus Livius*, that the 'citizens are to be watched so that they cannot under the cover of good do evil and so that they gain only such popularity as advances and does not harm liberty' – by which he meant the liberty of the state, see Skinner, *Machiavelli*, 75.

[3] Thomas Hobbes, *Leviathan* (Harmondsworth: Penguin, 1985).

European enlightenment out of the mists of history.[4] The unwarranted and often problematic assumptions of these older accounts of the state as a singular, mindful and sovereign agent have not been much curtailed by the dominant twentieth-century theories (from Weber and Foucault) that present the state as a dispersed agent of rationalisation, discipline or governmentality.[5] In each case the problems of what the state actually is, how it can be separated from society and how it acts have been neglected in favour of even larger questions about the operations of power on a cultural scale.

Some scholars have produced richly informative studies of the state by adopting simple definitions, or by neglecting them altogether. Using Weber's explanation that states were simply 'coercion wielding organisations' exercising unambiguous dominion over defined territories, Tilly was able to produce a summary explanation of the modern nation-state in Europe emerging from the combined demands of war-making and capital-raising.[6] While his study included the city-states of Italy and Germany, it was mostly unworried about the boundaries or relationships between the most powerful firms and the formal bureaucracies. In a similar way, Anderson was able to trace how the combined effects and demands of centralisation, empire-building, global trade, large-scale European war, bureaucratic innovation and reinvented forms of legal codification fashioned a club of absolutist monarchies in Europe between the fifteenth and twentieth centuries.[7] His history produced a distinctive description of the absolutist state – powerful, centralised, imperial, tax-raising arbiters of Roman law – without offering a theoretical definition of the state itself. Of course, early-modern politics was easier to understand in this respect. For Machiavelli, Hobbes and Hegel, and the absolutist states of early-modern Europe in general, the unambiguous presence of the sovereign monarch helped stabilise the concept of the state.

---

[4] Georg Wilhelm Friedrich Hegel, *Elements of the Philosophy of Right*, ed. A.W. Wood, trans. H.B. Nisbet (Cambridge University Press, 2008), 366–7; Enrique Dussel, 'Eurocentrism and Modernity (Introduction to the Frankfurt Lectures)', *Boundary 2* 20, no. 3 (1 October 1993): 71–4; M. Crawford Young, *The African Colonial State in Comparative Perspective* (New Haven: Yale University Press, 1994), 17–39.

[5] Max Weber, *Economy and Society*, ed. Guenther Roth and Claus Wittich, vol. 2 (Berkeley: University of California Press, 1978), 969–89; Michel Foucault, *Discipline and Punish: The Birth of the Prison* (London: Penguin, 1977); Michel Foucault, 'Governmentality', in *The Foucault Effect: Studies in Governmentality with Two Lectures by and an Interview with Michel Foucault*, ed. Peter Burchell, Colin Gordon and Peter Miller (London: Harvester Wheatsheaf, 1991), 87–104.

[6] Charles Tilly, *Coercion, Capital and European States, AD 990–1992* (Oxford: Blackwell, 1992), 26, 98, 110; Weber, *Economy and Society*, 54 for the original definition.

[7] Perry Anderson, *Lineages of the Absolutist State* (London: Verso, 1974).

In the twentieth century, and especially outside of the boundaries covered by the treaties of Westphalia, the problem is much more messy; on close examination the state has a disconcerting habit of unravelling to its constituent parts (or blurring seamlessly in to the corporations – like IBM – that now surround it).[8] The simplicity of the early-modern state seems fictional in this period. It was this imaginary unity that prompted the eminent structural functionalist, Radcliffe-Brown, in 1930 to ridicule the idea of the state as an entity with a will as 'a fiction of the philosophers' when he argued that 'there is no such thing as the power of the State; there are only, in reality, powers of individuals – kings, prime ministers, magistrates, policemen, party bosses, and voters'.[9] Many recent students of the state have taken this empirical scepticism about the unitary state further, adopting Abrams' observation that the idea of the state as an entity is a façade: 'an ideological artefact attributing unity, morality and independence to the disunited, amoral and dependent workings of the practice of government'.[10]

Anthropologists of the state have, accordingly, begun to explore the disparate ways in which the state is actually constituted – defined and experienced – in the languages that people use to describe and engage it. Gupta's anthropology of low-level officials in the villages of India, for example, shows convincingly that the rhetorics of corruption used by citizens, media and officials actually constitute the state and its effects.[11] In these accounts, far from being an autonomous institution, the state is a product of, and embedded within, the languages and practices of both the governed and the governing.

This idea of the state as a product of culture has been encouraged by a very influential turn across many disciplines towards the study of discourses – especially, recently, of science and engineering – in the formation of states in the nineteenth and twentieth centuries.[12] Foucault's

[8] James R. Beniger, *The Control Revolution: Technological and Economic Origins of the Information Society* (Cambridge, MA: Harvard University Press, 1986); Paul N. Edwards, *The Closed World: Computers and the Politics of Discourse in Cold War America* (Cambridge, MA: MIT Press, 1996).

[9] Alfred Radcliffe-Brown, 'Preface', in *African Political Systems*, ed. Edward Evan Evans-Pritchard, Meyer Fortes and Rachel Dempster (Oxford: International African Institute, 1955), xxiii, http://library.wur.nl/WebQuery/clc/464172.

[10] Philip Abrams, 'Notes on the Difficulty of Studying the State', *Journal of Historical Sociology* 1, no. 1 (1988): 81; Timothy Mitchell, 'Society, Economy, and the State Effect', in *State / Culture: State-Formation after the Cultural Turn*, ed. George Steinmetz (Ithaca: Cornell University Press, 1999). The same essay appears in Timothy Mitchell, 'Society, Economy and the State Effect', *The Anthropology of the State: A Reader* 9 (2006): 169.

[11] Akhil Gupta, 'Blurred Boundaries: The Discourse of Corruption, the Culture of Politics, and the Imagined State', *American Ethnologist* 22, no. 2 (1 May 1995): 375–402.

[12] The argument was made convincingly in Philip Richard D. Corrigan and Derek Sayer, *The Great Arch: English State Formation as Cultural Revolution* (Oxford: Blackwell, 1985);

key arguments about state power – first, of the ordering effects of the micro-disciplines of sciences and institutions in assembling the modern state[13] and, second, of the turn to a new kind of sovereignty (very different from Machiavelli's self-preserving state) derived from the biological well-being of populations[14] – have wrought something like a conceptual revolution. The history of the colonies has been important in this movement. Studies of experts' fashioning new forms and structures of power as they sought new kinds of knowledge on the African continent have been deeply illuminating.[15] But they have, also, tended to obscure the comparatively limited powers and scope of the colonial state, and the inadequacy of an account of the state (in Africa) motivated by the search for knowledge.

The state in Africa – both the colony and its successor – has been described by many historians as a gatekeeper. Cooper, in particular, has shown that colonial and post-colonial states on the African continent survived by standing 'astride the intersection of the colonial territory and the outside world' and, critically, that they 'had weak instruments for entering the social and cultural realm'.[16] Far from being driven by a ubiquitous scientific curiosity about the well-being of the population, African states were built in an informational void without the ability to 'track the individual body or understand the dynamics of the social body'.[17] Many important studies have discussed the forms of cheap indirect rule, blind

Khaled Fahmy, *All the Pasha's Men: Mehmed Ali, His Army, and the Making of Modern Egypt* (Cairo: American University in Cairo Press, 1997); Bernard S. Cohn, *Colonialism and Its Forms of Knowledge: The British in India* (Delhi: Oxford University Press, 1997); James C. Scott, *Seeing like a State: How Certain Schemes to Improve the Human Condition Have Failed* (New Haven: Yale University Press, 1998); Timothy Mitchell, *Rule of Experts: Egypt, Techno-Politics, Modernity* (Berkeley: University of California Press, 2002).

[13] Timothy Mitchell, 'The Limits of the State: Beyond Statist Approaches and Their Critics', *The American Political Science Review* (1991): 93–4; Foucault, *Discipline and Punish*.

[14] Foucault, 'Governmentality'; Mitchell, 'Society, Economy, and the State Effect', 87.

[15] Megan Vaughan, *Curing Their Ills: Colonial Power and African Illness* (Stanford: Stanford University Press, 1991); Scott, *Seeing like a State*; Mitchell, *Rule of Experts*; Patrick Harries, *Butterflies & Barbarians: Swiss Missionaries & Systems of Knowledge in South-East Africa* (London: James Currey Publishers, 2007); W. Beinart, *The Rise of Conservation in South Africa: Settlers, Livestock, and the Environment 1770–1950* (New York: Oxford University Press, 2008); William Beinart, Karen Brown and Daniel Gilfoyle, 'Experts and Expertise in Colonial Africa Reconsidered: Science and the Interpenetration of Knowledge', *African Affairs* 108, no. 432 (7 January 2009): 413–33; Helen Tilley, *Africa as a Living Laboratory: Empire, Development, and the Problem of Scientific Knowledge, 1870–1950* (University of Chicago Press, 2011). Each of these studies can be read to show the severe geographical and financial limits of the colonial state's power.

[16] Frederick Cooper, *Africa Since 1940: The Past of the Present* (Cambridge University Press, 2002), 5.

[17] Frederick Cooper, *Decolonization and African Society: The Labor Question in French and British Africa* (Cambridge University Press, 1996), 335.

tax-farming and skeletal physical and administrative infrastructure that characterised the colonial state.[18] These show that typically the state in Africa was built to control trade – it began at the harbour, expanded to form the colonial city, followed the line of rail and relied heavily on revenues from the export of a single commodity. This has – until very recently – changed little in the post-colonial era, with only the state's ability to 'defend the gate' undergoing significant variation. One of the objects of this book is to show that the South African state shares this gatekeeping architecture with its continental peers; another is that the same technologies of biometric registration that are now seen as the most promising remedy for bureaucratic incapacity on the African continent played an important part in limiting the colonial state's intellectual and administrative ambitions.

This very constrained scope means that a cultural understanding of the state – one which works well for the sprawling bureaucracies of India or France[19] – is problematic on the African continent. Nor is the cultural explanation of state power proposed by Bourdieu helped by the fact that the agents of what has in many respects proven to be hegemonic religious change – colonial missionaries of many kinds – were frequently bitterly at odds with the colonial state.[20] The state on the African continent has done little of the cultural work that historians have demonstrated in detail elsewhere,[21] and its grasp over the social, where it exists at all, has long been feeble.[22]

This means that two influential approaches to understanding the modern state – which we can call reification and anthropology – are both

---

[18] Walter Rodney, *How Europe Underdeveloped Africa* (Washington: Howard University Press, 1982), 208–9; Claude Ake, 'Rethinking African Democracy', *Journal of Democracy* 2, no. 1 (1991): 38; Sara Berry, *No Condition Is Permanent: The Social Dynamics of Agrarian Change in Sub-Saharan Africa* (Madison: University of Wisconsin Press, 1993), 22–42; Young, *The African Colonial State*, 100–29; Jeff Herbst, *States and Power in Africa: Comparative Lessons in Authority and Control* (Princeton University Press, 2000), 161–70.

[19] Gupta, 'Blurred Boundaries'; Pierre Bourdieu, 'Rethinking the State: Genesis and Structure of the Bureaucratic Field', in *State / Culture: State-Formation after the Cultural Turn*, ed. George Steinmetz (Ithaca: Cornell University Press, 1999).

[20] Jean Comaroff and John L. Comaroff, *Of Revelation and Revolution: The Dialectics of Modernity on a South African Frontier*, vol. 1 (University of Chicago Press, 1991), 261–87; Young, *The African Colonial State*, 97.

[21] Arno J. Mayer, *Persistence of the Old Regime: Europe to the Great War* (New York: Pantheon Books, 1981); P. Chatterjee, *Nationalist Thought and the Colonial World: A Derivative Discourse* (London: Zed Books, 1986); Corrigan and Sayer, *The Great Arch*; Alexander Woodside, *Lost Modernities: China, Vietnam, Korea, and the Hazards of World History* (Cambridge, MA: Harvard University Press, 2006).

[22] Jean-Francois Bayart, *The State in Africa: The Politics of the Belly* (London: Longman, 1993).

ill-suited to the project I have in mind here. Fortunately there is another materialist and technological understanding of the state which becomes evident from a careful consideration of the premises of the sociology of the state. In his 1977 discussion 'on the difficult of studying the state' Abrams was concerned to demonstrate that the state is an ideological artefact; to do this he drew on an observation in Miliband's study of the capitalist state which has since become famous: 'the "state" is not a thing ... it does not, as such, exist'.[23] Yet in an important sense Abrams (and Miliband) are clearly both wrong. For the state is very much a thing, or, perhaps more accurately, it is a constellation of things: roads, hospitals, telecommunication lines, computers, filing cabinets, weapons, bullion (to name only a few). And at the core of these things is a coordinating bureaucracy which, as many scholars have argued over many years, is very much an object itself. 'The very term bureaucracy points to a form of governance built around a thing', Hull notes in his work on the technologies of the Pakistani state, 'the writing desk, and the documentary practices it supports'.[24] Historians have examined the physical work of the bureaucracy in detail over many centuries and the results are compelling: at the heart of the state lies a literary and paper-processing machine, one which has been identified, studied and described in careful detail.[25] Indeed it was this literary project of government – as Clanchy, Goody, Corrigan and Sayer, Gorski, and Hull have each shown – that provided the vehicle for the cultural changes that were produced by the modern state.[26]

Following the paperwork, of course, can confront the same problems that Radcliffe-Brown identified for the state as an assembly of people. How do we distinguish the writings of the church, or the records of the largest corporations, from the formal bureaucracy? I think that this can be very illuminating – Clanchy's work on the medieval English state is one excellent example, and Hull's recent study of the Pakistani state is another – by reconstructing the pathways that paper follows in the

---

[23] Abrams, 'Notes on the Difficulty of Studying the State', 69.

[24] Matthew S. Hull, 'Ruled by Records: The Expropriation of Land and the Misappropriation of Lists in Islamabad', *American Ethnologist* 35, no. 4 (2008): 501–18.

[25] M.T. Clanchy, *From Memory to Written Record, England 1066–1307* (London: Edward Arnold, 1979); Corrigan and Sayer, *The Great Arch*; Beniger, *The Control Revolution*; Jack Goody, *The Logic of Writing and the Organization of Society* (Oxford University Press, 1986); P. Sankar, 'State Power and Record-Keeping: The History of Individualized Surveillance in the United States, 1790–1935' (University of Pennsylvania, 1992); Valentin Groebner, *Who Are You? Identification, Deception, and Surveillance in Early Modern Europe* (New York: Zone Books, 2007); Jon Agar, *The Government Machine: A Revolutionary History of the Computer* (Cambridge, MA: MIT Press, 2003).

[26] Corrigan and Sayer, *The Great Arch*; Philip S. Gorski, 'The Protestant Ethic Revisited: Disciplinary Revolution and State Formation in Holland and Prussia', *The American Journal of Sociology* 99, no. 2 (1993): 265–316; Hull, 'Ruled by Records'.

bureaucracy;[27] over the last two centuries that would often mean tracking files as they move between parliament, the executive branches of the government and officials in the provinces.[28] But in this book I will not be doing that in part because the state I am examining explicitly disavowed the old, dispersed documentary basis of bureaucracy. I have another set of paper-handling conventions in mind.

If, as Weber observed, the processing of paperwork is the defining characteristic of modern bureaucracy in both its public and private forms,[29] a particular subset of those documents – technologies of identification – lies at the heart of the work that the state arrogates to itself. These acts of identification can take many forms – passports, identity cards, birth certificates, drivers' licences – but they all, typically, hinge on processes of civil registration, the official recording of identification at birth, marriage and death.[30] Processes of identification working together make up an infrastructure of citizenship – a set of slowly emerging rules, standards and networks of communication – which give any state distinctive centres and, as many of the protagonists in this story insisted, a distinctive political character.[31] When I speak of the biometric state I have in mind a state that is organised around technologies and architectures of identification that are very different – and which function politically very differently – from the older forms of written identification that have produced the modern state.

## A new state

The documentary state is very old. Its key elements – the registration of property, of tax and military recruitment liabilities and the recording of

---

[27] Clanchy, *From Memory to Written Record*; Matthew S. Hull, 'Documents and Bureaucracy', *Annual Review of Anthropology* 41 (2012): 251–67.

[28] An excellent example is Geoffrey Parker, *The Grand Strategy of Philip II* (New Haven: Yale University Press, 1998) which builds a study of the Spanish empire around the movement of paper.

[29] Weber, *Economy and Society*, 957–88.

[30] Jane Caplan and John C. Torpey, *Documenting Individual Identity: The Development of State Practices in the Modern World* (Princeton University Press, 2001); John Torpey, *The Invention of the Passport: Surveillance, Citizenship and the State* (Cambridge University Press, 2000); Colin J. Bennett and David Lyon, *Playing the Identity Card: Surveillance, Security and Identification in Global Perspective*, 1st edn (New York: Routledge, 2008); Keith Breckenridge and Simon Szreter, eds, *Registration and Recognition: Documenting the Person in World History*, Proceedings of the British Academy 182 (Oxford University Press, 2012).

[31] Simon Szreter and Keith Breckenridge, 'Recognition and Registration: The Infrastructure of Personhood in World History', in *Registration and Recognition: Documenting the Person in World History*, ed. Keith Breckenridge and Simon Szreter, Proceedings of the British Academy 182 (Oxford University Press, 2012), 1–36.

personal and family names – have existed for thousands of years in the rice-growing societies of Asia.[32] In Europe these familiar features were formed a little more recently, mainly between the eleventh and the fourteenth centuries. Over 300 years, as Clanchy has particularly shown for England, writing, blessed by its association with an ascendant church, fitfully usurped the status and claims of oral and iconic forms of authority and power. In practice this meant that parchment documents (often forged by church officials) replaced spoken claims as guarantors of property and propriety; writing became the basis of law, and the main instrument of state extractions like taxation and recruitment; a new class of literate officials leaked from the church into the royal chanceries and then spread out – as agents of central government – to the parishes in the countryside.[33] Over the next half-millenium written record making and keeping became a massive and dense field of culture, acting to preserve and simplify property and to discipline the poor. This may have been preeminently the case in England, as Corrigan and Sayer have suggested, but historians have traced the administrative powers of writing in very similar processes throughout Europe, the Americas and parts of Asia.[34] It is no wonder then that the powers of documentary government rest (typically undisturbed by rude empirical enquiry) at the heart of the most influential theories of state power produced in the century that separates the writings of Max Weber and James Scott.[35]

[32] Richard von Glahn, 'Household Registration, Property Rights, and Social Obligations in Imperial China: Principles and Practices', in *Registration and Recognition: Documenting the Person in World History*, ed. Keith Breckenridge and Simon Szreter, Proceedings of the British Academy 182 (Oxford University Press, 2012), 39–66; James C. Scott, *The Art of Not Being Governed: An Anarchist History of Upland Southeast Asia* (New Haven: Yale University Press, 2009); Woodside, *Lost Modernities*; on the Ancient Near East see Goody, *The Logic of Writing*.

[33] Clanchy, *From Memory to Written Record*; for similar process in Europe, see Groebner, *Who Are You?*; on the persistence of spoken and communal forms of respectability in Spain and Spanish America, see T. Herzog, *Defining Nations: Immigrants and Citizens in Early Modern Spain and Spanish America* (New Haven: Yale University Press, 2003).

[34] Jane Caplan, '"This or That Particular Person": Protocols of Identification in Nineteenth-Century Europe', in *Documenting Individual Identity: The Development of State Practices in the Modern World*, ed. Jane Caplan and John Torpey (Princeton University Press, 2001), 49–66; Corrigan and Sayer, *The Great Arch*; Torpey, *The Invention of the Passport*; Gorski, 'The Protestant Ethic Revisited'; Jill Lepore, *The Name of War: King Philip's War and the Origins of American Identity* (New York: Knopf, 1998); Walter D. Mignolo, *Darker Side of the Renaissance: Literacy, Territoriality and Colonization* (Ann Arbor: University of Michigan, 1995); Parker, *The Grand Strategy of Philip II*; Sankar, 'State Power and Record-Keeping'; Michael Warner, *The Letters of the Republic: Publication and the Public Sphere in Eighteenth Century America* (Cambridge, MA: Harvard University Press, 1990).

[35] Weber, *Economy and Society*, vol. 2, 957–94; Foucault, *Discipline and Punish*, especially 184–96; A. Giddens, *The Nation-State and Violence*, vol. 2, *A Contemporary Critique of Historical Materialism* (Berkeley: University of California Press, 1985), 174–96; Michael

In our own time a transformation very like the one that Clanchy described seems to be under way. Some parts of this process are well known. Since the early 1970s, a globally networked, digital order – in which the most important information processing systems are outsourced to, or owned by, one of a small group of international corporations – has come to dominate most of the planet. There is no novelty in this claim; many important writers have pointed to elements of the process over the last two decades.

Twenty years ago Sassen showed that a global city had emerged from the real-time trading in financial markets in London, New York and Tokyo. The citizens of this global city, often connected to each other by computer terminals, continue to live mostly detached from the levelling constraints of local states (even after they have been rescued from bankruptcy by taxpayer bailouts).[36] 'Today, we live in a global society', Mann observed before it had become obvious: 'It is not a unitary society, nor is it an ideological community or a state, but it is a single power network.'[37] In a similar vein, Castells followed the influence of transnational firms, multilateral institutions and tightly organised global economies in the fashioning of a twenty-first century network state. (Castells has been proven wrong about Africa being structurally excluded from this network society – where something like the opposite has actually happened – but he was not alone in this.)[38]

In the richest countries Lyon has traced a new kind of surveillance state emerging from the feedback and storage capabilities of ubiquitous computers and the twin imperatives of controlling integrated welfare services and global national security.[39] Ironically these grand informational ambitions seem actually to have weakened many of the old surveillance and managerial powers of the documentary state. Agar, following the administrative and information-handling capacity of the British state in detail over the twentieth century, has shown that the contradictory imperatives to manage almost universal welfare benefits and reduce costs through the deployment of large-scale computer systems after the 1970s

Mann, *The Sources of Social Power: The Rise of Classes and Nation-States, 1760–1914*, vol. 2 (Cambridge University Press, 1993), 40–2, 282–5, stresses communication and education.

[36] Saskia Sassen, *The Global City: New York, London, Tokyo* (Princeton University Press, 1991).

[37] Mann, *The Sources of Social Power*, vol. 2, 11.

[38] Manuel Castells, *The Rise of the Network Society* (Oxford: Blackwell, 1996), 88–102; Manuel Castells, *The Power of Identity*, vol. 2, 2nd edn (Oxford: Wiley Blackwell, 1997), 303–366.

[39] D. Lyon, *The Electronic Eye: The Rise of Surveillance Society* (Minneapolis: University of Minnesota Press, 1994), 83–118; Agar, *The Government Machine*, 369–77.

has produced a much weakened and hollowed-out state, one in which officials have only the vaguest idea how the work of information processing is actually done. The network state lies in the hands of a cluster of overlapping information technology companies. Some, like IBM, have a history of supporting the information processing requirements of the documentary state that date back a century, but a shifting host of intrinsically global firms provide database and transactional services that are well beyond the capacities of even the most skilled officials.[40] This new state is geographically and institutionally very different from the documentary order that Clanchy described, and it is also very unlike the expert (and omnipotent) bureaucracy that Weber saw as the revolutionary agent of rationalization.

We have grown accustomed to the idea – perhaps since Geertz first observed that 'the culture of a people is an ensemble of texts' – that reading (and writing) are indispensable for the assembling of knowledge. The analytical primacy of literacy is, on the one hand, a powerful hermeneutic method, allowing us, as Geertz suggested, 'to read over the shoulders' of our subjects.[41] Literacy has also, as Scott's studies of the politics of legibility have amply shown,[42] structured the ancient and modern states' effort to know and govern their subjects. Yet understanding the significance of the biometric state requires us to break with the convention that writing and legibility – what Foucault called the 'power of writing'[43]– provide an adequate account of technologies of information, and the new forms of administrative power. To understand this point requires us to consider the meanings and nature of biometric technologies very closely.

## Biometrics

What do we mean when we use the word biometric? The word refers to two distinct but closely related subjects. For most of the twentieth century biometrics referred to the statistical science of biological data analysis, and particularly to the mathematical methods – the correlation coefficient, regression, the goodness of fit test and many other techniques – that equipped statistics with analytical and predictive powers that have fostered its supremacy in biology, economics, finance and

[40] Edwards, *The Closed World*.

[41] Clifford Geertz, *The Interpretation of Cultures: Selected Essays* (New York: Basic Books, 1973), 452; Christopher Norris, *Derrida* (London: Fontana, 1987).

[42] Scott, *Seeing like a State*, discusses legibility, all too briefly, on 68–9; Scott, *The Art of Not Being Governed*, argues that people on the mountainous fringes of documentary states have resisted literacy in order to protect their independence.

[43] Foucault, *Discipline and Punish*, 189.

many fields of medicine. Remarkably, between 1900 and the middle of the 1960s, the epicentre of this statistical revolution was the little Galton Laboratory at the University College, London, a unit that was itself the combination of two earlier centres founded by Francis Galton: the Eugenics and Biometrics Laboratories. Throughout this period the science of biometrics retained an intimate and increasingly fraught relationship with the eugenics movement, but it was essentially a theoretical and mathematical science.[44] It was only at the very end of the last century that the statistical science of biometrics began to face an identity crisis of its own.

At its 1998 meetings held in South Africa, the president of the International Biometrics Society announced that the organisation had registered the word biometrics as a trademark in an effort to resist its use in the 'popular media' as a description of the new 'techniques being developed for identification of individuals'.[45] The attempt failed. In the years since, commercial interest in technologies of identification has almost drowned out the older scientific meanings of the word; the Biometrics Society's trademark, for example, is overwhelmed by nearly 200 competing trademarks that use the term biometrics, and thousands that include some element of biometric identification in their descriptions.

This new, and now very popular, idea of biometric identification refers to the automated recognition of individuals based on precisely measured features of the body. The concept of computerised identification, like many other technologies of our era, existed in the domain of science fiction long before it had a name or existed in practice: the oldest reference to computerised voice recognition, for example, is in Kubrick's 1968 film *2001: A Space Odyssey*. Yet the technical term – biometrics – for computerised identification only came in to use in the late 1970s.[46] It was clearly the proliferation of computers, of optical scanners, and the networks that connect them, that was integral to the appropriation and popularisation of the word. Biometrics, in this sense, can best be described as the identification of people by machines.

[44] D.A. MacKenzie, *Statistics in Britain, 1865–1930: The Social Construction of Scientific Knowledge* (Edinburgh: Edinburgh University Press, 1981), 101–49; D.J. Kevles, *In the Name of Eugenics: Genetics and the Uses of Human Heredity* (Cambridge, MA: Harvard University Press, 1995), 222.

[45] Susan R. Wilson, 'Evolution and Biometry', *Biometrics* 55, no. 2 (June 1999): 334.

[46] Andrew Pollack, 'Recognizing the Real You', *New York Times*, 24 September 1981; Louis Katz, 'Biometric Measuring Device' (New York, 15 May 1979). Here I disagree with Caplan and with Fraenkel that the signature or the photograph, like the fingerprint, serve as distinctively individualising signs. What distinguishes the fingerprint in this new popular usage of biometrics (and in my own understanding) is the ability to extract standardised mathematical data from each contact. Fraenkel's argument is explained in Caplan, 'Protocols of Identification in Nineteenth-Century Europe', 51–3.

The roots of these technologies clearly lie in anthropometry – the meticulous systems of measurement of the body that were developed by the French police in the nineteenth century. Computerised biometric systems – like face recognition and hand geometry – have adopted, quite directly, many of the older mechanical tools of measurement that were developed by the Parisian police administrator, Alphonse Bertillon, in the 1880s.[47] But it has been fingerprinting, more than any other bodily measurement, that has nurtured the development of automated identification. The formalisation of the technology of fingerprinting, as many people know, was Francis Galton's work, and the focus of his considerable energies between 1889 and 1901.

It is important to remember that Galton's interest in fingerprinting preceded the formal definition of biometric statistics as a field of knowledge, and that his work on fingerprinting as identification prompted the key inventions of statistics: this was his effort to highlight the dangers of 'co-relation' in Bertillon's mathematics of identification. Galton showed that Bertillon's use of the measurement of body parts as randomly selected numbers to generate large numerical claims to uniqueness was mistaken – the length of the arm was directly related to the length of the ear – and in the process he invented the idea of correlation, one of the key methods of modern statistics. But Galton was always very interested in the practical, what we might call the technological, benefits of fingerprinting, in policing, prisons and banking and particularly (as I show in the chapters below) in their application to the problems of governing the empire.

There is, then, some historical justice in the fact that, over the last decade, the technology of fingerprinting has usurped the rights of the science of biological statistics to their common name. But it is also important to see that the political relationship between these two siblings – one science, the other technology – is not an amicable one. When the scholars at the US National Research Council recently noted 'the curiosity that two fields so linked in Galton's work should a century later have few points of contact', they worried that the separation was motivated by the fact that biometric identification 'is scientifically less basic'. But the reason for the separation is probably more banal: throughout the twentieth century Galton's biometric identification remained preoccupied with the often brutal practical tasks of policing, which, as Cole has shown, allowed little

---

[47] Allan Sekula, 'The Body and the Archive', *October* 39 (1986): 3–64; Carlo Ginzburg, 'Clues: Roots of an Evidential Paradigm', in *Clues, Myths and the Historical Method* (Baltimore: Johns Hopkins University Press, 1989); Sankar, 'State Power and Record-Keeping', 155–89; Simon A. Cole, *Suspect Identities: A History of Fingerprinting and Criminal Identification* (Cambridge, MA: Harvard University Press, 2001), 15–81.

space for scholarly scepticism and the implications of probability theory for the burden of proof in the courtroom. This, as I explain in the chapters below, was especially the case in the territories of the former empire where Galton's technologies developed unhindered by a liberal legal order.[48] Yet, because they share a common founder, fingerprinting and statistics share many important characteristics.

Biometrics, in both scientific and technological forms, are intrinsically mathematical entities. This seems odd because we are accustomed to thinking of them as printed products of the body. Galton (like many other biometric advocates) was fond of describing fingerprinting as a mimetic text, as 'self-signatures' where 'the hand of the accused person prints its own impression'.[49] He liked to compare fingerprint classifications with words. 'A set of finger prints', he wrote, 'may be so described by a few letters, that it can be easily searched for and found in any large collection, just as the name of a person is found in a directory'. And it is true that one of the key features of fingerprints – before the 1980s – was that they were retained in the archives (like letters and photographs) long after they had done the original work of identification.[50] But the practice of biometric identification is itself an intrinsically statistical activity with very little in common with reading.

These administrative biometrics are numerical representations of patterns on the human body. They may, initially, be derived from images – usually of fingerprints, sometimes of irises or faces – but they are always transformed through the extraction of patterns and minutiae points in to a very large number that will support a claim for uniqueness in the human population.[51] Although the work is usually done by computer

[48] Whither Biometrics Committee, National Research Council, *Biometric Recognition: Challenges and Opportunities*, ed. Joseph N. Pato and Lynette I. Millett (Washington, DC: The National Academies Press, 2010), 17; Cole, *Suspect Identities*, 199–216.

[49] Francis Galton, *Finger Prints* (London and New York: Macmillan and Co., 1892), 168; Karl Pearson, *The Life, Letters and Labours of Francis Galton: Correlation, Personal Identification and Eugenics*, vol. 3A (Cambridge University Press, 1930), 154.

[50] Cole, *Suspect Identities*, 87; Sekula, 'The Body and the Archive', 15–16.

[51] It is also true that biometrics is a product of the effort to join statistics with portraiture in the nineteenth century. It emerged from the effort to satisfy what Pinney has called the 'indexical yearning', the pervasive ambition to find an undeniable documentary technique for capturing a human subject. A continuity runs from phrenology, through anthropometrics. Textualisation is key to these arguments. Biometrics, even more than photographs, are indexical signs because they are usually made through physical contact with the subject, they are – in C.S. Peirce's sense – contiguous with their subject. C. Pinney, *Camera Indica: The Social Life of Indian Photographs* (University of Chicago Press, 1997), 14. An indexical continuity between phrenology, anthropometry and biometics has been important in all the key works on this subject to date. Sekula, 'The Body and the Archive'; Ginzburg, 'Clues'; Pinney, *Camera Indica*. This mimetic quality to biometrics is important politically, as I have already said, but it tends to obscure the very

sensors, the method for extracting these very large distinguishing numbers for biometrics has changed remarkably little since Francis Galton
first described it in 1891.[52]

In the chapters that follow I show (unlike those who have written on
this subject to date) that for much of its century-long history biometric
administration has been self-consciously opposed to documentary government. This is a departure from the theoretical work of some of the
most important studies of fingerprinting, and it must be developed carefully.[53] Two basic points can be made here. From the first plans for the
introduction of fingerprinting that were drawn up by Galton, biometric
administration was motivated by a desire to identify the illiterate subjects
of Britain's imperial possessions.[54] Remarkably this project – of fixing the
names of illiterate African subjects in particular – remained the driving
justification through the whole of the twentieth century and it is still the
*raison d'être* of the current round of large-scale biometric systems, both in
the former colonies and at the gates of the imperial capitals.

Another difference is material. While the roots of fingerprinting lie in
the nineteenth-century effort to create a 'link between an individual body
and a paper record', biometrics are not documents and the databases
that retain them are not archives in any meaningful sense of that word.[55]
These modern biometric identifiers typically exist only intangibly, stored

---

different delinguistic mathematical indexing that makes fingerprinting work. In fact the
workings of fingerprint identification required a process of pure mathematical abstraction, both for the tiny points on a single fingerprint (which Galton dubbed minutiae),
or for the string used to identify a set of ten carefully taken prints in a collection of tens
or hundreds of thousands. This was a fairly straightforward matter of probability, but in
practice it proved very difficult to arrange. Finding a workable mathematical method of
classifying tens of thousands of prints – the key to his plans to fingerprint everyone – was
something that eluded Galton. The solution was developed in India, in the offices of
the Bengal Police by some, still uncertain, combination of the Commissioner, Edward
Henry, and two of his subordinates, Azizul Haque and Hem Chandra Bose. Chandak
Sengoopta, *Imprint of the Raj: How Fingerprinting Was Born in Colonial India* (London:
Macmillan, 2003), 138–46. Mathematical simplification and abstraction was important to large-scale systems of fingerprinting, but after the work was done by computer
it became all inclusive. Biometrics was technology driven by numerical, probabilistic
abstraction – free of writing.

[52] Francis Galton, 'Identification by Finger Tips', *Nineteenth Century* 30 (August
1891): 307.

[53] Cole, *Suspect Identities*; Sankar, 'State Power and Record-Keeping'; Sekula, 'The Body
and the Archive'. But I agree with Higgs, that the politics of biometrics lies in a deeply
entrenched status distinction in English history between identification by paper and the
marking of the body, see Edward Higgs, 'Fingerprints and Citizenship: The British State
and the Identification of Pensioners in the Interwar Period', *History Workshop Journal* 69
(2010): 52–67.

[54] Francis Galton, 'Identification Offices in India and Egypt', *Nineteenth Century* 48 (July
1900): 118–26; Galton, *Finger Prints*, 27, 149–50.

[55] Cole, *Suspect Identities*, 4, 14–82 on the emergence of fingerprint classification.

in a database or written in to the memory of an integrated-circuit on a smart-card. Now, to be clear, biometric tools have sometimes served to supplement the existing systems of documentary government – to close the gap, for example, between a passport and its bearer. But they have also, and much more commonly, been used to curtail or obliterate an existing (and often inadequate) system of documentary government. An effort to escape the limits of the old paper state – of slow, susceptible or unreliable bureaucratic processing, of forgery, deception and translation in the preparation of documents – lies at the core of the effort to develop biometric identification technologies. And this political imperative – to sweep away the slow and messy and unreliable paper-based systems of government – remains a key part of the appeal of these systems.

In this book I want to draw attention to the peculiar geography of the new biometric state. But it is probably worth pointing out, first, that universal biometric registration can fairly be described as the *bête noire* of scholarly and popular fears of the overweening surveillance state; these fears have been eloquently captured in Andrew Niccol's 1997 popular dystopian film *Gattaca* and in Giorgio Agamben's bitter denunciation of biometrics as the apex of an intrinsically genocidal liberal order.[56] The mathematical efficiency of the biometric state brings with it possibilities for storing and processing data, and for generating feedback about the behaviour of individuals that was simply unmanageable in a paper bureaucracy. This brings us much closer to the all-knowing cybernetic state that Norbert Wiener predicted long ago, and it gives a chilling edge to Habermas' worries about the steering effects of the 'technicizing of the lifeworld'.[57] How this all works in practice remains to be seen, but it is, I think, fair to say that both Wiener and Habermas would be surprised that this technologically precocious state is taking form outside of the developed West. This book is an explanation of how that has happened.

Biometric identification systems are under development in many regions and institutions around the world. The new passport documents in Europe, North America and Australia all make use of biometrics,

---

[56] Giorgio Agamben, *State of Exception* (University of Chicago Press, 2005); Giorgio Agamben, *Homo Sacer: Sovereign Power and Bare Life* (Berkeley: University of California Press, 1998); Malcolm Bull, 'States Don't Really Mind Their Citizens Dying (Provided They Don't All Do It at Once): They Just Don't Like Anyone Else to Kill Them', *London Review of Books* 26, no. 24 (16 December 2004): 3–6.

[57] Wiener's worry was expressed in the words of critic writing in *Le Monde*. Norbert Wiener, *The Human Use of Human Beings: Cybernetics and Society* (Cambridge, MA: Da Capo Press, 1988); J. Habermas, *The Theory of Communicative Action. Vol. 2: Lifeworld and System: A Critique of Functionalist Reason* (Boston: Beacon Press, 1992), especially 181–5.

Countries with biometric civil and voting registration schemes

but they have very limited surveillance capacities because – under the bright light of popular anxiety about bureaucratic invasions of privacy – they have been deliberated and carefully hobbled. In stark contrast, foreign migrants in these same countries have been subjected to much more powerful ten fingerprint and iris-capturing systems that are centrally gathered, and shared amongst all of the signatory states of the Treaty of Schengen.[58] There are some obvious imperial legacies in the identification, and policing, of these target populations. But it is still incongruous, in the light of the wider scholarship on the new surveillance state, that the most powerful biometric surveillance systems are being developed in the poorest countries, the former colonies of the European empires.[59]

Biometric civil registration systems – often linked to the payment of cash transfers for the poor – are currently under development in Brazil and Mexico, and in dozens of other countries. But the most ambitious – and

---

[58] J.P. Aus, *Decision-Making under Pressure: The Negotiation of the Biometric Passports Regulation in the Council* (ARENA Working Paper, 11, 2006); J.P. Aus, 'Eurodac: A Solution Looking for a Problem?', *European Integration Online Papers (EIoP)* 10, no. 6 (2006).

[59] The northern emphasis in the scholarship of the surveillance state is implicit in David Lyon, *Surveillance after September 11* (Malden, MA: Polity Press in association with Blackwell Pub. Inc., 2003); and explicit in Colin J. Bennett and Charles D. Raab, *The Governance of Privacy: Policy Instruments in Global Perspective* (Aldershot: Ashgate, 2003). These biometric states closely resemble the territory Castells described as falling outside of the global information economy. Castells, *The Rise of the Network Society*, 133–6, 359; see especially Manuel Castells, *End of Millenium*, vol. 3 (Oxford: Blackwell, 1998), 82–128. The South African technological inheritance for the continent is much less liberating than he may have anticipated.

the most important – of all of these schemes is the Aadhaar project in contemporary India, a breathtaking instance of social engineering which plans to register the identities of hundreds of millions of people over the next decade. The project is closely associated with Nandan Nilekani, co-founder of Infosys, one of the most successful of the Indian back-office outsourcing firms, and author of the progressive manifesto *Imagining India*.[60] Aadhaar has roots in the same cluster of national security, broken civil registration, banking and welfare transfer systems that have motivated biometric population registration elsewhere. Under the influence of Infosys's experience supporting computerised record-keeping, it presents a very pure example of biometric registration as the antithesis of documentary bureaucracy. Citizens provide digitised images of their fingerprints and their irises and they are issued, in return, only with an identifying number called the UID – the scheme provides no official identity card, no entitlements and, formally, no document proving registration. (This unreal quality is supplemented by the dubious legal status of the scheme which, despite having registered nearly 200 million people by the end of 2013, had no statutory basis.) The UID – a technical term for the super index that supports relationships between tables in modern database design – will function as a digital lynchpin for a host of other bureaucratic transactions, but its virtual, numerical, privatised and radically centralised character distinguish it from the customary work of bureaucracy everywhere, and from the long-established practice of the Indian state in particular.[61]

Nilekani's UID project highlights the immaterial, mathematical goals of biometric registration, and it is an amazing test case of the capabilities of the biometric state – nothing remotely similar in scale has ever been attempted. But it is also important in the global history of biometrics because it marks the restoration of the trajectory of biometric government to India after its century-long confinement in South Africa. Fingerprinting lies at the heart of the fraught relationship between these two countries, a relationship that began with the struggle between Smuts and Gandhi, and which continued to shape the United Nations and the system of international relations in the second half of the last century.[62]

[60] N. Nilekani, *Imagining India: Ideas for the New Century* (London: Allen Lane, 2008).
[61] I am grateful to have heard the papers and discussion at the CSDS/Sarai Workshop on the Social Life of Information, Delhi, 14–16 November 2013. See www.facebook.com/sarai.net.
[62] Mark Mazower, *No Enchanted Palace: The End of Empire and the Ideological Origins of the United Nations*, Lawrence Stone Lectures (Princeton University Press, 2009); Newell M. Stultz, 'Evolution of the United Nations Anti-Apartheid Regime', *Human Rights Quarterly* 13, no. 1 (February 1, 1991): 1–23.

Biometric government – as many scholars have shown – was first developed in India, it was brought to South Africa by officials of the Indian Colonial Service in 1900, where it was quickly put to use against the Indians in the Transvaal. It was Gandhi's international protests against fingerprint registration in South Africa that prompted the development of his anti-colonial political philosophy. These same protests killed off the development of the large-scale fingerprint identification schemes in India. In South Africa their development continued, unconstrained. In broad outline the events of this history are well known – because Gandhi worked hard to communicate them. Yet the details are still surprising and important, not least because Gandhi produced a sophisticated political analysis of the different forms of biometric registration, and their moral effects, while he was in South Africa. That assessment has, to date, had intriguingly little influence on the debates around Aadhaar.

### The South African story

The history that follows is not a political narrative or an intellectual history in the conventional sense. My interest is in the form of the state, not the causes of Apartheid, or the policies of the African National Congress. The explanation takes the form of a sequence of episodes, starting with Galton's journey to South Africa in 1850 and ending, over a century later, with the development of a system of biometric cash transfers in the KwaZulu Bantustan. In each generation between these events the state attempted an elaborate scheme of biometric registration and my chapters follow those projects. Failures and unintended outcomes have been such important parts of each of these schemes that my reader might be tempted to view the twentieth-century history of South Africa as the failure of the biometric state – indeed, in some respects, that is an insightful characterisation of the institutional and political problems that exist in this country. But the inter-generational struggle to establish large-scale biometric registration systems has also endowed South Africa with what economists call a distinctive path dependency, in this case one that has been anchored around the institutions of identification. Where identification in almost every other society has emerged from the demands of local government (especially at the level of the village, the parish or the municipality), in South Africa it can be undertaken only by a single, central government agency (based in Pretoria), and only by means of fingerprinting. This biometric centralisation has been in place for half a century, and it affects almost every aspect of institutional life in South Africa – from banking to vehicle licensing. The arrangement (as I show in the last chapter) is globally distinctive, but very few people – inside the country or outside of it – are aware of that fact, or of its consequences. Nor is there a similar awareness, outside of South Africa, that biometric registration may imply the adoption of this peculiar architecture of government.

Like several of the recent studies of the global systems of migration control, this book shows that South Africa had a special place as a twentieth-century laboratory of empire, for the technologies of racial segregation, and – as Hannah Arendt first suggested – for the marriage of bureaucracy and despotism. It was, as these studies show, both a site for the evolution of precocious forms of bureaucratically arranged racial supremacy, and a very important stage of resistance to racism addressed to a global public.[63] The global origins and effects of this state have been studied for a long time. There is also an older historiography that shows the direct links between the forms of segregation that developed in South Africa and progressive segregationism in the American South in the 1930s and 1940s.[64] And both Mazower and Mitchell have recently shown – separately – that the global politics of the arrangements for segregation in South Africa had powerful effects on international institutions in the twentieth century.[65] In this book I want to turn again to Arendt's argument (much of which is wrong in fact although right in principle) that South Africa was the 'culture-bed of Imperialism' and the fulcrum of distinctive forms of racialised bureaucracy that have spread around the world.[66]

---

[63] Adam McKeown, *Melancholy Order: Asian Migration and the Globalization of Borders* (New York: Columbia University Press, 2008); M. Lake and H. Reynolds, *Drawing the Global Colour Line: White Men's Countries and the International Challenge of Racial Equality* (Cambridge University Press, 2008). Arendt places South Africa at the centre of her history of totalitarian government, and she is very interested in the intersections of racism and bureaucracy. Her arguments, especially about the timing and consequences of the segregationist order in South Africa for European totalitarianism, are very provocative, but it is difficult to know how one might actually test them. Unfortunately she says far too little on bureaucracy and far too much, mostly utterly speculative, about racism and anti-Semitism and their effects on the mob. Hannah Arendt, *The Origins of Totalitarianism* (San Diego: Harcourt Brace and Company, 1973), 185–221.

[64] C. Vann Woodward, *The Strange Career of Jim Crow* (New York: Oxford University Press, 1966); John W. Cell, *The Highest Stage of White Supremacy: The Origins of Segregation in South Africa and the American South* (Cambridge University Press, 1982); A. Lichtenstein, 'Good Roads and Chain Gangs in the Progressive South: "The Negro Convict Is a Slave"', *The Journal of Southern History* 59, no. 1 (1993): 85–110. Interestingly, the processes that Massey and Denton have famously described as American Apartheid actually bear very little resemblance to practices of urban segregation in South Africa, see: D.S. Massey and N.A. Denton, *American Apartheid* (Cambridge, MA: Harvard University Press, 1994).

[65] Mazower, *No Enchanted Palace*; Timothy Mitchell, *Carbon Democracy: Political Power in the Age of Oil* (London: Verso, 2011), 72–83.

[66] Arendt, *The Origins of Totalitarianism*, 185–221.

In the chapters that remain I will show that for a century the South African state has served as a laboratory for this new form of biometric government, and that the technologies that states across the world have been adopting over the last decade find their earliest and fullest development here. It is important to be precise about this. The biometric systems of this new kind of government are complex and confounding, involving international networks of ideas, tools, firms and states – this is as true today as it was a century ago. Latour's insistence that the technology itself is an actor in these networks – setting constraints, possibilities and failures independently – also applies here.[67] Biometric administration has been global from its origins (for a discussion of this see the Epilogue) with key sites of development in India, Argentina, England and France, and the United States of America. Yet, in each case, the plans of biometric social engineers have been undone before they could approach the scale and embeddedness that has been achieved in South Africa.

I begin the story with an exploration of the significance of the two meanings of the word biometric. The first, and until quite recently, the most important, refers to the very large and influential science of biological statistics, the second to the technology that uses physical characteristics of the body for identification. Both fields trace their origins to the work of Francis Galton. In the different fields that explore the history of statistics and the history of surveillance, Galton is typically treated as a figure of European intellectual history. Standing between Bertillon or Quetelet and Edward Henry or J. Edgar Hoover, Galton's political preoccupations have usually been described as metropolitan in focus. In Chapter 1 I show that Galton should more properly be seen as an archetypical imperial intellectual, long before Karl Pearson's announcement in 1900 of the search for a 'new anthropology' that could guide the progressive imperial state.[68] Galton was an African, and especially a South African, expert in the half-century before the South African War. Most importantly Galton used the racial insights from his travels in South Africa as the evidence for the emerging statistical science of eugenics. Long before he had any usable evidence from his anthropometric laboratory Galton had derived the key claims about the implications of the normal curve for human descent using his South African evidence. His views on Africa were fiercely derogatory, and, like Carlyle, he publicly and repeatedly

---

[67] Bruno Latour, *Reassembling the Social: An Introduction to Actor-Network-Theory* (New York: Oxford University Press, 2005).

[68] Karl Pearson, *National Life from the Standpoint of Science* (London: A & C Black, 1900), 84, www.archive.org/details/nationallifefrom00pearrich; G.R. Searle, *The Quest for National Efficiency: A Study in British Politics and Political Thought, 1899–1914* (Oxford: Basil Blackwell, 1971), 36–40.

rejected the humanitarian critique of slavery, arguing for coercive forms of labour mobilisation on the continent because he believed that black people were suited to slavery. Most importantly Galton's South African writing was used as evidence to turn his cousin, Charles Darwin, away from humanitarianism to the pessimistic politics of eugenics, effectively overthrowing the political trajectory of the older man's entire life's work. By arguing that individuals were trapped in hereditary racial populations that would ineluctably revert to a statistical average, Galton provided the argument that would be used, especially by Lionel Curtis, to build the case for the segregationist state. In a letter written to the magazine *Nineteenth Century* he was also the first person publicly to recommend the use of large-scale fingerprinting in South Africa.

Joseph Chamberlain sent Edward Henry to South Africa in July 1900 to establish a new Criminal Investigation Department in Johannesburg, and, in the words of the dispatch announcing his appointment, 'introduce a scientific system of identification'. As I show in Chapter 2 it was Henry working alone in Johannesburg, in close personal correspondence with Sir Alfred Milner in Cape Town, who managed the transition from the Boer state's controls of the movement of Africans to the new order. Henry also set up the staffing and regulations of the Transvaal Town Police, many months before the Transvaal was actually safe enough to support a civil police force. He drafted the new manual of police practice – with its special concern for the fact that 'natives are known by detachable names', the Police Act, and personally hired the first group of recruits. Clearly working from the evidence and report of a seminal 1896 Mining Industry Commission report that was dominated by the key American engineers in the Transvaal, Henry set up the characteristic divisions of the South African Police – the Liquor Branch, Gold Branch and the CID – and he directed the Town Police to enforce the old Republican Pass Law, all to meet the special needs of the mining industry. Henry worked industriously between August 1900 and March 1901, when he returned to London to take up a position as heir apparent in the Metropolitan Police. Many historians have commented on his importance in the development of the twentieth-century Scotland Yard, but his role in South Africa has gone largely unnoticed.

Wherever Henry went in South Africa fingerprint repositories sprouted like mushrooms; he changed the late 1890s' fledgling interest in Bertillonage to a systematic determination to gather fingerprints, using his classification system, in the Transvaal Town Police, the Native Affairs Department and individual mines; he founded similar repositories in Natal, the Cape and the Orange River Colony. Some measure of his importance to the new Reconstruction regime can be gathered from the fact that Milner tried, unsuccessfully, to appoint him as the Commissioner of Police. The system of pass controls that Henry built on the Rand during Reconstruction, with funds raised from the wages of hundreds of thousands of mine workers, was more intrusive and more long-lasting than any similar fingerprinting regime on the planet. And in the years after his departure the managers of the different repositories were the key agents of the project, eventually only realised under Verwoerd in the 1950s, to centralise and synchronise the different fingerprint registries. Yet for much of the period between 1910 and 1940 the state was reluctant to adopt fingerprinting to manage the control of all Africans in South Africa, and the reasons for that lie largely with the actions of a gentle Indian lawyer who arrived in South Africa in 1893.

The story of Gandhi's adoption of the politics of satyagraha in response to Smuts' efforts to impose the 'racial taint' of fingerprint registration on all Asian immigrants to the Transvaal is probably the most well-known episode in this story. Inevitably the detail of events in South Africa is more complicated, and paradoxical, than the story presented in the biographies of the Mahatma, or the histories that rely on them. Chapter 3 re-examines the segregationist plans, laid by the imperial officials, Milner and Lionel Curtis, for a ruthless system of registration designed to 'shut the gate' to prevent Indian immigration to the Transvaal. But it also shows that, before Curtis' scheme was developed, Gandhi had himself proposed that fingerprints should be included as one of the requirements of the new law 'to regulate the signing of negotiable instruments by Indians' in Natal in 1904. The target of Gandhi's worries about the regulation of fraudulent contracts was the illiterate indentured workers in Natal. Gradually he realised that compulsory full-print fingerprinting of all Indians, even the very literate, formed the core of Curtis' segregationist plan. In building popular opposition to registration after 1906, Gandhi articulated a very powerful case that the imposition of fingerprinting represented a violation of the sanctity of the Indian family, and, in particular, of the masculine honour of Indian men. It was this sentimental argument, built in outraged dialogue with the partially internalised criticisms of the white press, that sustained the popular opposition to fingerprinting. But it also made it very difficult for Gandhi to give effect to his startling proposals for mass voluntary registration in January 1908.

In the effort to win back his constituency Gandhi effectively became, in the early months of 1908, an advocate of the scientific and progressive virtues of fingerprinting. He mastered Edward Henry's work, and became an astute critic of the politics of the different systems of fingerprint registration. (Many of his points about the moral and political difference between one-to-one and one-to-many identification techniques apply to the systems being developed today.) The philosophy of satya-graha and the anti-progressive politics of *Hind Swaraj* emerged, in part, out of his contradictory involvement in the fingerprinting policy. Gandhi addressed the slight to masculine honour by arguing that satyagraha required an exalted manhood and extreme forms of courage. His distinctively non-national anti-colonialism derived from the argument that Hindus and Muslims (including the peoples of the Middle East) were 'sons of the same Mother India'. But, in his elaboration of this manly resistance, Gandhi massively overstated its power, particularly the resisters' ability to withdraw their consent at any later time. In fact the biometric registrations of early 1907 could never be withdrawn and the Indians of the Transvaal remained subjected to a comprehensive system of racial control until the end of the 1960s.

In the most important sense Gandhi lost the battle with the Transvaal state, leaving his constituents subject to an invasive biographical archive that regulated their property rights, movement and even the members of their families. But the effects of his struggle were strongly felt for a full generation after 1908. For the first three decades of the new Union of South Africa, mining officials, doctors and policemen argued repeatedly for the introduction of a system of universal fingerprint identification. Often they used the example of the fingerprint register that had been built to control the identification, payment and policing of the 60,000 Chinese labourers that had been brought to South Africa during Milner's government. But these efforts were all resisted by the key decision-makers. Some of the efforts at centralisation were undone by technical limits on the amalgamation of fingerprint archives, and some by the parochial interests of local officials. But the most important influence was clearly that officials learned, after Gandhi, to be suspicious of the political dangers of the overweening claims of the advocates of fingerprint registration. By the end of the 1930s that lesson had been forgotten, and the state began to turn, once again, to the old project of universal registration. In the intervening years, the advocates of biometrics managed to undermine an existing system of civil registration amongst black South Africans in the countryside.

For many scholars of empire and the modern state it is an article of faith that a will to know – a compelling desire for comprehensive and universal information – has motivated the development of administrative systems, and the relationship between states and society, since the middle of the nineteenth century. But in South Africa, and in colonial Africa more generally, there is scarcely any evidence of this kind of administrative curiosity. Government in Africa, which scholars have variously described as a gatekeeper state, as decentralised despotism and as hegemony on a shoestring, has been defined much more by the absence of information than its presence. Several scholars have traced the model for the system of indirect rule that dominated much of British Africa to the old colony of Natal. In Chapter 4 I examine the rise and fall of a system of birth registration for Africans in Natal. The chapter shows that vital registration (recording births and deaths) was implemented by officials seconded to Natal from the Indian Colonial Service in the early years of the twentieth century. Over the course of time, key police officials, who were also the most enthusiastic advocates of biometric registration, built a case against the system on the grounds that educated Africans could not be trusted to provide truthful vital information. Under pressure from the advocates of fingerprinting, and from the accountants, the scheme was abandoned in the early 1920s. Over the next three decades, African leaders and public health officials lobbied constantly for the restoration of the system of civil registration. Only in the late 1940s, as the new Apartheid bureaucracy began to take shape, did the state turn again to the plan of compulsory rural vital registration. And this time compulsory fingerprinting, which obviated the question of who black citizens thought they were, was key to the project. Civil registration, under Apartheid, was in any case undone by the state's efforts to enforce a fingerprint identity document on all African men and women.

Hendrik Verwoerd imposed the Bewysburostelsel – the bureau of proof regime – on South Africa during the 1950s as part of the effort to build a race-based Population Register. In Chapter 5 I show that the Bewysburo was a grand effort – proposed by English-speaking officials – finally to sweep away the many different, paper-based, systems of control, taxation and identification that had been imposed in piecemeal fashion on the African population of South Africa in the half-century after Milner. Every African man and, for the first time, every woman was issued with a personal identification card attached to a Reference Book (Bewysboek) that maintained a complete history of employment, residence and taxation. The issuing of these Bewysboeke coincided with the building of centralised registers of this information, and, critically, the collection of a single centralised collection of fingerprints. Fingerprinting was, by deliberate design, the only mechanism available to the bureaucracy to establish the integrity of the new documents, and by extension the Africans they were intended to regulate. Chapter 5 is a narrative of the administrative disorder that followed from the building of this central biometric population register for all Africans, the issuing of identity cards and the effort to classify the huge body of fingerprints that poured in from the countryside. It examines internally generated crises and some of the ways those subjected to the Bewysburo sought to defeat it. By the late 1950s the Bewysburo fingerprinting scheme had failed to meet even the most basic goals of its original designers, leaving in place ubiquitous violence and an international symbol of the totalitarian character of the white state. But the Bewysburo project also placed South Africa on a path of biometric government that it could not abandon.

Chapter 6 takes up what is perhaps the most curious part of this story: the transformation of a coercive arrangement of centralised biometric surveillance to a new kind of welfare state, characterised by universally distributed cash payments that are delivered biometrically. By the early 1980s fingerprinting had become part of the national security strategy of an increasingly militarised government. The decision to extend finger-printing to white people followed a series of attacks on oil installations by white members of *Umkhonto we Sizwe*, the African National Congress's armed wing. This new policy coincided with the announcement of the computerised technologies of biometric identification. Towards the end of the 1980s a very precocious system of biometric pension delivery was developed in the KwaZulu and Kangwane homelands, under pressure from feminist human rights organisations inside the country. It was this system which allowed the democratic state to expand the pool of welfare recipients in 2002; from an original population of some two million – mostly pension – recipients the grant-receiving population now num-bers more than fourteen million people, many of them young women. This system is delivered by privately held, financial service providers, and dominated by Net1 UEPS, a firm with a global ambition to displace Visa and MasterCard amongst the world's poor.

In the Epilogue I take up the problem of comparing the South African history with the two other societies – the United States and Argentina – that have the most fully developed preoccupations with government-by-fingerprinting in the twentieth century. The chapter considers two related problems of comparative method which demand a serious discussion of similar processes in these two countries: the first asks why biometric registration, although applied in many other countries, never approached South African forms of centralisation or scope; the second asks what it was that motivated biometric government, both in general and specif-ically in the South African case. Here, in place of the widely used gen-eral explanations of state-building as a product of governmentality and rationalisation offered by Foucault and Weber, I point to the very specific history of progressivism, and to its distinctively unconstrained role in the making of the South African state after 1900. Readers looking for my theoretical inclinations might want to start there.

# 1  Science of empire: the South African origins and objects of Galtonian eugenics

Biometric technologies are mathematical tools of identification. They rely for both their basic workings and, much less obviously, for a new set of economic functions on the operations of modern statistics, and especially on probabilities. Today, the fine-grained, individualised feedback data that is generated by biometric transactions is especially valuable to banks and other credit providers, and serves as one of the most powerful drivers of biometric identity registration around the world.[1] This chapter examines the curious coincidence that both techniques – the statistical tools of inference, correlation and regression and the technologies of fingerprinting – can be traced directly to Francis Galton, the fountainhead of modern statistics and the inventor of fingerprinting. The explanation broaches another little known fact: that Galton was an African, and especially a South African, scientific expert. The global politics of the South African history of biometric government began with Galton's journey to the region half-a-century before the new state was formed.

Until quite recently our scholarship has been inclined to view Africa as a territory outside of science, one that has been powerfully shaped by colonial mastery of technologies like the Gatling gun and quinine, but which, outside of a few, very specific fields – like paleoanthropology – has been largely excluded from the global scientific conversation.[2] In recent years researchers have been interrogating this idea, focusing on twentieth-century efforts to indigenise science on the continent, especially through the efforts of the South African prime minister, Jan Smuts, and his followers.[3] In the pages that follow, I retrace a similar process,

---

[1] Keith Breckenridge, 'The World's First Biometric Money: Ghana's E-Zwich and the Contemporary Influence of South African Biometrics', *Africa: The Journal of the International African Institute* 80, no. 4 (2010): 642–62.

[2] D.R. Headrick, *The Tools of Empire: Technology and European Imperialism in the Nineteenth Century* (Oxford University Press, 1981); Michael Adas, *Machines as the Measure of Men: Science, Technology and Ideologies of Western Dominance* (Ithaca: Cornell University Press, 1989).

[3] See, especially, Saul Dubow, *A Commonwealth of Knowledge: Science, Sensibility, and White South Africa, 1820–2000* (Oxford and New York: Oxford University Press, 2006); Helen

but one that has had very destructive political effects: the African origins of what can be described as the Galtonian revolution – the application of the normal distribution curve (and its effects) to almost every aspect of human society. The chapter traces two, related but paradoxical, circles that have both been largely neglected. The first is that the numerical science that Galton triggered in the last quarter of the nineteenth century had important origins in Southern Africa, ironically fostering an intellectual movement which effectively defined Africans as living outside of science. The second is that Galton's South African expertise played an important part in the emergence of Social Darwinism, and, especially, in the conversion of Darwin himself to the pessimistic biology of eugenics in the 1870s. These arguments also returned to South Africa – carried mainly through Herbert Spencer's popular anthropologies – fashioning the intellectual world of late nineteenth-century imperialism, and creating a political theory that supported the coercive forms of segregation that were distinctive of the South African state. The most important of these institutions and regulations relied, in turn, on the fingerprinting technologies that Galton developed in the 1880s.

## Galton's world

We live in a world governed by metrics of every sort. Globally standardised numerical scales measure our intelligence, emotional aptitudes, creditworthiness, and our performance at play and work. This is especially true of the world of commerce. Our economies – whether they make pork-bellies, silicon chips or mortgages – are driven by new kinds of global markets formed by interchangeable numerical indices and statistical tools of analysis.[4] Popular assessments of what we should eat, and

Tilley, *Africa as a Living Laboratory: Empire, Development, and the Problem of Scientific Knowledge, 1870–1950* (University of Chicago Press, 2011). Other studies that have looked at Southern Africa as an important site in the global production of science include Elizabeth Green Musselman, 'Swords into Ploughshares: John Herschel's Progressive View of Astronomical and Imperial Governance', *The British Journal for the History of Science* 31, no. 4 (1 December 1998): 419–35; Patrick Harries, *Butterflies & Barbarians: Swiss Missionaries & Systems of Knowledge in South-East Africa* (London: James Currey Publishers, 2007); W. Beinart, *The Rise of Conservation in South Africa: Settlers, Livestock, and the Environment 1770–1950* (New York: Oxford University Press, 2008), 28–63, describes the adoption of indigenous skills by travellers, hunters and even the English settlers in the first half of the nineteenth century. On Africans as makers and users of new technologies see Clapperton Mavhunga, 'Firearms Diffusion, Exotic and Indigenous Knowledge Systems in the Lowveld Frontier, South Eastern Zimbabwe 1870–1920', *Comparative Technology Transfer and Society* 1, no. 2 (2003): 201–31.

[4] D.A. MacKenzie, *An Engine, Not a Camera: How Financial Models Shape Markets* (Cambridge, MA: MIT Press, 2006).

what medicines we should not use, are determined by the ubiquitous vocabulary of statistics and the very expensive surveys that generate those numbers. Even the meaning of the word for evidence – which for centuries implied personal experience and oral testimony – now refers to the most abstracted forms of statistical knowledge, and, in its most exulted form, to the double-blind, randomised control-trial. This regime of numbers is Francis Galton's world.

The universal trust in abstracted numbers is new, in many fields dating only from the 1970s. Standardised systems of creating and sharing measurements emerged, as Theodore Porter shows, almost by accident, with many of the professions that have become masters of the universe of metrics – the accountants, engineers and physicians – resisting their development tooth and nail. The ascendancy of statistics was fostered, across many fields and regions, by the breakdown of face-to-face systems of trust. It was the state – and the regulation of the professions under the often fraught and contested politics of the democracies – that drove the development of standardised, rule-bound, numerical forms of expert knowledge. Statistics implied disinterested, objective knowledge.[5] If the pursuit of an abstracted, mechanical objectivity across all areas of institutional life was politically motivated, we should still ask – given their ubiquitous power – what it is about statistics that appeals to those who use them. Aside from the banal claim that science requires measurement (which can be easily satisfied) what is it about statistical ways of knowing that appeals to their advocates? These virtues of metrics are as old as modern statistics, which is to say not very old at all – they date back to the second half of the nineteenth century, and to the development of the field of biometrics by Francis Galton and his brilliant disciple Karl Pearson.[6] The statistical sciences that Galton and Pearson conceived and developed were motivated by a cluster of aesthetic virtues that have remained with us. First amongst these was abstraction. Statisticians, from the middle of the nineteenth century, began to use and manipulate single measurements – like height – across entire populations which allowed them to easily order and understand large populations. This use of a carefully selected small set of numerical measures to stand for very complex social or economic processes remains at the heart of modern

[5] Theodore M. Porter, *Trust in Numbers: The Pursuit of Objectivity in Science and Public Life* (Princeton University Press, 1995).
[6] For an account of the development of modern statistical methods, see Ian Hacking, *The Taming of Chance* (Cambridge University Press, 1990); S.M. Stigler, *The History of Statistics: The Measurement of Uncertainty before 1900* (New York: Belknap Press, 1986) also credits Edgeworth and Yule.

statistics (and a fertile source of controversy[7]). And the reason of course is that numerical abstraction supports (and often requires) intense forms of simplification. By gathering, and generalising, a single measure – like Gross National Product or Intelligence Quotient or Fatality-free Shifts or Price – researchers (and managers) are able to slough off a host of other, potentially confusing, characteristics. Intense forms of abstraction and generalisation also encourage powerfully asymmetrical forms of knowledge, with one individual or institution able, for the first time, to measure very large populations. These virtues increased the scope of what could be researched but they were also very well matched to a new global infrastructure of knowledge, one in which computers and their networks served as both producers and consumers of very large amounts of information.

Mechanical forms of calculation were important to statistics from the beginning – Galton, like many of the Victorians, was obsessed with devices that could standardise and measure independently of personal assessment. From 1894 Pearson was famously attached to a Brunsviga 'brain of steel' calculating machine, and he predicted (and longed for) a time when machines would make light of the exhausting labour of mathematical calculations.[8] That moment finally arrived in the early 1970s, but it has grown apace in the decades since. The global proliferation of computers, their networks and their astonishing programmable capacities for both simplifying calculation routines and generating standardised, numerical feedback has produced a corresponding expansion in the scope and power of statistics. This new mechanised world of numerical analysis involves – indeed it requires – an escape from language, and, especially, freedom from the exhausting responsibility for reading.[9]

Numerical indices – unlike problems in the real world – have the wonderful advantage of being intrinsically and obviously tractable: they can be moved, easily. School attendance or achievement scores – unlike the presence of real begging children on the street – typically respond to state investments. (This is especially true where those gathering the data are encouraged to make fine adjustments to their selection criteria by monetary rewards for performance that are derived from the same scores.) But the real allure of modern statistics comes from the powers of the new

[7] Morten Jerven, *Poor Numbers: How We Are Misled by African Development Statistics and What to Do about It* (Ithaca: Cornell University Press, 2013).

[8] Theodore M. Porter, *Karl Pearson: The Scientific Life in a Statistical Age* (Princeton University Press, 2004), 241.

[9] Pearson, to be fair, read (and wrote) prodigiously throughout his life, as his five-volume biography of Galton demonstrates, but see also his earlier obsession with German history and feminism beautifully described in Porter, *Karl Pearson*.

measures of probability, measures that Galton derived from the effects of
the normal distribution. The first of these was correlation, which allows
researchers to make claims about the dependent relationships between
metrics (the graphical correspondence, for example, between heights
and the lengths of noses). The second was regression – the ranking pro-
cedure that Galton used to show that in biological populations future
generations will tend to revert to an accumulated historical average.
Regression effectively produced a new philosophical account of causality, an 'imperialism of probabilities',[10] expanding over the course of the
twentieth century far away from the normal distribution and Galton's
obsession with biological populations. Statistical inference, Pearson's
special contribution, is the third magical power of modern statistics: a
measure of the ability to make claims about large unknown populations
by drawing, again, on the probabilistic qualities of the normal distribution. In combination, regression and inference have allowed researchers
and institutions to make powerful, and often accurate, predictions about
future events. And that ability – working in combination with the generalisation of standardised measures – has changed the world.

Remarkably, for most of the twentieth century a single, small, academic institution acted as the incubator of what has amounted to a global
civilisational change. This was the Galton Laboratory at the University
of London which was the product of the union, in 1904, of the Eugenics
Record Office and the Biometric Laboratory. Under Pearson and his suc-
cessors the poorly resourced lab became the engine of modern statistical
methods – developing and applying ever more flexible and sophisticated
techniques to a host of problems. This is a story that is well known even if
it is not as widely acknowledged in our social theory as it should be.[11]

This book tells the story of another laboratory, of a society that served
as the incubator of forms of biometric government that were first planned
and advocated by Galton. Those forms of biometric registration and identification are bringing the world's poor within the grasp of statistical forms
of government for the first time. A key difference in these territories – in
part because of the norms of colonial rule and in part because of the
severity of poverty – is that privacy exists very weakly in law and in practice. The new forms of statistical creditworthiness will be bound directly
to older forms of biometric identification. This may bring in to being, in

---

[10] Hacking, *The Taming of Chance*, 5.

[11] D.A. MacKenzie, *Statistics in Britain, 1865–1930: The Social Construction of Scientific
Knowledge* (Edinburgh: Edinburgh University Press, 1981), 105–9; Stigler, *The History
of Statistics*, 361; D.J. Kevles, *In the Name of Eugenics: Genetics and the Uses of Human
Heredity* (Cambridge, MA: Harvard University Press, 1995), 222; Hacking, *The Taming
of Chance*, 182.

the former colonial world, a much more severe and invasive Galtonian order than our current one. There is an interesting circularity at work here, for many of Galton's ideas had their origins in Southern Africa.

Who, then, was Francis Galton? He was, like his older half-cousin, Charles Darwin, the grandson of the late eighteenth-century Enlightenment iconoclast, Erasmus Darwin.[12] In the 1840s, before he travelled to South Africa, he was a delinquent member of the extended tribe of wealthy non-conformist and abolitionist intellectuals who effectively monopolised the intellectual life of Britain in the first half of the nineteenth century. This first decade of Galton's adult life was a study in upper-middle-class under-achievement. He had tried his hand as a medical student, and attempted the mathematical tripos at Cambridge (which would later be his key test of genius) but failed at both. In 1844 his father died, leaving him independently wealthy and with little reason to indulge the habit of frenzied work that he would later argue was the distinguishing characteristic of the civilised races. For the next six years he indulged himself, touring Egypt and Syria and hunting on the Scottish estates of wealthy friends. And it was hunting that encouraged him at the end of the decade to 'travel in South Africa, which had a potent attraction for those who wished to combine the joy of exploration with that of encountering big game'.[13] His life changed dramatically after he returned from South Africa in 1852. In the second half of the nineteenth century Galton became famous as the author of two distinct fields of knowledge, both of which we now name biometrics.

The first of these is the science of empirical statistics that concerns itself with living things, and, during Galton's life, with the biology of human beings. For decades Galton's statistics was motivated by what he argued was a decline 'of the better sorts' but his complaints met with little popular or political success. That changed at the turn of the twentieth century as the early catastrophes of the South African War combined with the evidence from Booth's surveys of poverty in London to prompt British panic about physical deterioration. Galton became the celebrated champion of a popular eugenics movement that lasted until the First World War.[14]

---

[12] Adrian J. Desmond and James Richard Moore, *Darwin's Sacred Cause: How a Hatred of Slavery Shaped Darwin's Views on Human Evolution* (Boston: Houghton Mifflin Harcourt, 2009), 17.

[13] Sir Francis Galton, *Memories of My Life* (London: Methuen & Co., 1908), 124.

[14] MacKenzie, *Statistics in Britain*; Kevles, *In the Name of Eugenics*, 57; Simon Szreter, *Fertility, Class and Gender in Britain, 1860–1940* (Cambridge University Press, 2002), 96–148.

The scholarship on Galtonian statistics is broad, complex and very interesting, but it can be summarised. Statistics was a widely practised amateur science in the nineteenth century, and Galton was only one of many researchers drawn to the virtues of numerical ways of describing and understanding the physical and – especially from the middle of the century – the social sciences. He was not an especially gifted or diligent mathematician, but he was pivotal and revolutionary. For it was Galton who adopted the theory of what was already well known as the error curve – the inverted bell jar graph we know as the normal distribution – and overturned its political meaning. He reversed the political logic of Quetelet's interest in the Aristotelian man, turning normality into mediocrity. It was Galton's obsession with ranking, and the outliers of behaviour and ability, that produced his theory of regression to the mean and his explanation of 'co-relation' – two of the master procedures of contemporary statistics.[15]

In the accounts that scholars have offered of this statistical history, Galton's interest in Africa was important in establishing his prestige, but it was otherwise irrelevant. In the studies of Galton's writings after 1850 there has been a deliberate narrowing of the scope of the political terrain of his life.[16] Many of these scholars studiously ignore Galton's writing on Africa, and their political significance. This is because the history of Galton's writing has been organised by the problems that have retained their significance today – his interest in the normal curve gives way naturally to the coefficients of regression and correlation. This metropolitan preoccupation is pithily expressed by Stigler, the very eminent historian of statistics, who observed that it 'seems impossible to deny that Darwin's theories opened an intellectual continent more promising and attractive to the explorer Galton than Africa had been'.[17]

Galton has long occupied a special place in the social history of eugenics – the middle-class effort to discipline and control the fertility of the urban poor in Britain. Here the political goals of his statistics and his policy recommendations have been closely connected to the the Victorian middle class' reaction to rising working-class democracy

---

[15] Theodore M. Porter, *The Rise of Statistical Thinking 1820–1900* (Princeton University Press, 1986); Hacking, *The Taming of Chance*; Stigler, *The History of Statistics*; MacKenzie, *Statistics in Britain*.

[16] Raymond E. Fancher, 'Francis Galton's African Ethnography and Its Role in the Development of His Psychology', *The British Journal for the History of Science* 16, no. 1 (March 1983): 67–79 is an important exception; N.W. Gillham, *A Life of Sir Francis Galton: From African Exploration to the Birth of Eugenics* (New York: Oxford University Press, 2001) highlights his African experience without exploring its role in his statistics.

[17] Stigler, *The History of Statistics*, 267.

from the 1860s.[18] Imperial politics was carefully explored in the emergence after 1900 of the National Efficiency crisis, which saw Galton and Pearson, and many other important social-imperialists, arguing for unflinching genocide in Africa and the adoption of hard eugenic policies at home.[19] But the gap between the 1860s, when Galton's writing on eugenics began, and the early years of the twentieth century is curious. In this period, with the exception of his work in the early 1880s on prison photography, Galton was primarily concerned with gathering biographical and anthropometric data from English elites.[20] This is because between 1850 and 1900 Galton's primary juxtaposition was between the English elite and Africans, especially the Herero pastoralists he encountered in South Africa. This polite drawing of a veil over Galton's writings on Africa is true even of the very recent studies of eugenics in the colonial world.[21]

To be fair the connections between Galton's travels in South Africa and his interest in eugenics have been noticed by several historians, but most have moved quickly on, embarrassed, perhaps, by the extremity of his views on Africans. Nancy Stepan suggested that his journey was of 'prime importance' in the development of the views on race he expressed after *Hereditary Genius*, but her explanation of the place of Africans in Galton's theory is made necessarily brief by the scope of her study. Her analysis of the conceptual consequences of his work is also truncated. She concludes, I think too hastily, that Galton was working with an imprecise and ambiguous concept of race.[22] Stocking, similarly, argued that Galton's African experience led him towards a 'pessimistic

---

[18] G.S. Jones, *Outcast London: A Study in the Relationship between the Classes in Victorian Society* (New York: Pantheon, 1971); Anna Davin, 'Imperialism and Motherhood', *History Workshop Journal* 5 (1978): 9–65; MacKenzie, *Statistics in Britain*; Szreter, *Fertility, Class and Gender in Britain*; R.J. Overy, *The Morbid Age: Britain between the Wars* (London: Allen Lane, 2009).

[19] Bernard Semmel, *Imperialism and Social Reform: English Social-Imperial Thought, 1895–1914* (London: George Allen & Unwin, 1960), 41; G.R. Searle, *The Quest for National Efficiency: A Study in British Politics and Political Thought, 1899–1914* (Oxford: Basil Blackwell, 1971); Szreter, *Fertility, Class and Gender in Britain*, 148.

[20] Noted in Szreter, *Fertility, Class and Gender in Britain*, 142; Ruth Schwartz Cowan, 'Francis Galton's Statistical Ideas: The Influence of Eugenics', *Isis* 63, no. 4 (1972): 509–28. Galton's use of the word residuum, which after 1900 came to refer to the irredeemable poor, was applied to a very different population in this earlier period: 'the residuum that forms the bulk of the general society of small provincial places, is commonly very pure in its mediocrity', quoted in Karl Pearson, *The Life, Letters and Labours of Francis Galton: Researches of Middle Life*, vol. 2 (Cambridge University Press, 1924), 91; Jones, *Outcast London*, 330–1.

[21] Philippa Levine, 'Anthropology, Colonialism, and Eugenics', in *The Oxford Handbook of the History of Eugenics*, 2010, 43–61.

[22] Nancy Stepan, *The Idea of Race in Science: Great Britain, 1800–1960* (Basingstoke: Macmillan in association with St Antony's College Oxford, 1982), 126, 129.

view of civilization in which biological mechanisms were centrally problematic', but he did not pause to explain how this happened.[23] Raymond
Fancher, the historian of psychology, has looked most carefully at the
travel writing. He notes that Galton 'seems to have gone out of his way
to believe and report the worst' about the people he was observing and
that his observations were 'almost always less restrained and fairminded
than the parallel reports of his second-in-command, Charles Andersson'.
And he concludes that 'unflattering depictions of the African's character
and intellect formed important parts of Galton's arguments in both of
his seminal works'.[24] Again, however, the reader is left wondering how
exactly the African descriptions worked.

A very similar explanatory structure works in the now extensive historiography on the technologies of identification and registration. For
Galton was also the key figure behind the other field which we now call
biometrics, and particularly the methods and procedures of the technology of fingerprinting which are both the oldest and the most extensive forms of biometric identification.[25] In the history of policing, and
state identification systems, Galton is renowned as the fountainhead of
the most important practices of involuntary identification.[26] In these histories he is situated in an intellectual genealogy that connects Jeremy
Bentham with Alphonse Bertillon, working as a clerk for the Paris Police
in the 1870s. It was Bertillon who solved the problems of linking the
human body to the written register by applying anthropometrics – the
statistics of the body – to the process of identifying 'incurable vagrants',
building a tool that allowed the police to follow the criminal 'across time'
by indexing the body itself.[27] It was Bertillon's interest in the statistics of
probability that established the practical basis of biometry by specifying
in minute detail the procedures that should be used to measure, describe
and record eleven different parts of the body. Bertillonage, as the global system of criminal identification was called in the 1890s, injected
the measuring tools and racist preoccupations of phrenology and crani-
ometry, the fields of anthropology that sought to assess personality and

---

[23] George W. Stocking, *Victorian Anthropology* (New York: Free Press, 1991), 96.
[24] Fancher, 'Francis Galton's African Ethnography', 72, 79.
[25] See the essays in Jane Caplan and John C. Torpey, *Documenting Individual Identity:
The Development of State Practices in the Modern World* (Princeton University Press,
2001).
[26] Simon A. Cole, *Suspect Identities: A History of Fingerprinting and Criminal Identification*
(Cambridge, MA: Harvard University Press, 2001); P. Sankar, 'State Power and Record-
Keeping: The History of Individualized Surveillance in the United States, 1790–1935'
(University of Pennsylvania, 1992); Chandak Sengoopta, *Imprint of the Raj: How
Fingerprinting Was Born in Colonial India* (London: Macmillan, 2003).
[27] Cole, *Suspect Identities*, 33, 48.

intellect by measuring the contours, size and shape of the skull, into the heart of the modern bureaucracy. Galton's work on fingerprinting, in these accounts, was a logical remedy to the hidden statistical weaknesses of Bertillon's anthropometrics.

In recent years researchers have opened up the workings of the laboratory of courts and prisons in nineteenth-century India, showing, persuasively, that the original questions and the final practical solutions to the problems of large-scale registration emerged from the demands, and the resources, of colonial government.[28] Typically, however, the African history slips in to these accounts in a sentence or a paragraph noting that Edward Henry visited the Witwatersrand in 1900, that Gandhi became famous fighting Smuts over compulsory fingerprinting, or that Apartheid was the 'last physiognomic system of domination in the world'.[29] The problem is not easily solved by examining Galton from the other side of the world. Historians of Africa have returned the favour by presuming the significance of Galton's travels. In Jan-Bart Gewald's history of the Herero people there is a single sentence, in footnote 1 no less, that explains succinctly and with no further elaboration: 'Galton used his expedition experiences to develop the theory of eugenics.'[30] In his comprehensive intellectual history of race in South Africa, Dubow observed that Galton won scholarly prestige from 'his accounts as a pioneer explorer in South West Africa (Namibia) between 1850 and 1852' and that 'these experiences helped to shape Galton's racial attitudes and considerably affected his later formulation of eugenics'.[31] But, because of the enduring power of anatomical racism in South Africa, Dubow devotes his considerable

[28] Radhika Singha, 'Settle, Mobilize, Verify: Identification Practices in Colonial India', *Studies in History* 16, no. 2 (2000): 151–98; Cole, *Suspect Identities*; Sengoopta, *Imprint of the Raj*; Clare Anderson, *Legible Bodies: Race, Criminality, and Colonialism in South Asia* (Oxford and New York: Berg, 2004).

[29] Carlo Ginzburg, 'Clues: Roots of an Evidential Paradigm', in *Clues, Myths and the Historical Method* (Baltimore: Johns Hopkins University Press, 1989) 123; Allan Sekula, 'The Body and the Archive', *October* 39 (1986): 63; Edward Higgs, *Identifying the English: A History of Personal Identification, 1500 to the Present* (London and New York: Continuum, 2011).

[30] Jan-Bart Gewald, *Herero Heroes: A Socio-Political History of the Herero of Namibia, 1890–1923* (Athens, OH: Ohio State University Press, 1999), 10.

[31] Saul Dubow, *Illicit Union: Scientific Racism in Modern South Africa* (Johannesburg: Witwatersrand University Press, 1995), 15. Others who have noticed Galton's journey include Johannes Fabian, 'Hindsight: Thoughts on Anthropology Upon Reading Francis Galton's Narrative of an Explorer in Tropical South Africa (1853)', *Critique of Anthropology* 7, no. 2 (1987): 37–49; and Robert J. Gordon, 'The Venal Hottentot Venus and the Great Chain of Being', *African Studies* 51, no. 2 (1992): 185–201. Both are particularly interested in the infamous incident of Galton's description of the Hottentot Venus. G. Steinmetz, *The Devil's Handwriting: Precoloniality and the German Colonial State in Qingdao, Samoa, and Southwest Africa* (University of Chicago Press, 2007) notes Galton's place in the evolution of the ethnography of the Herero without linking it to the larger, and very important, argument.

energies to another explorer – Robert Knox, Galton's contemporary and opponent. My object here is to reinsert Galton's African experience into the development of eugenics, but also to show that Galton's two biometrics – eugenically motivated statistics of human biology and the technologies of fingerprint identification – have common roots in the project of Empire.

The two areas of Galtonian biometrics – statistics and identification – have evolved, and they have been studied, as separate fields. The results are perplexing. In a recent report for the US National Research Council, which effectively marked the death-knell of the project of biometric national identification for American citizens, the most eminent scientists working in the field of biometric identification lamented the absence of a probabilistic and statistical understanding of the claims of identification systems.[32] 'In distinguishing our topic from biometrics in its biostatistical sense', the report's author observed, 'one must note the curiosity that two fields so linked in Galton's work should a century later have few points of contact'. Yet, viewed from outside of the metropolis, the reason for the separation between the abstract sciences of statistics and the messy and brutal technology of fingerprinting is almost self-evident. Imperial government was the link between Galton's statistics and his technologies of identification.

## Humanitarian upheaval

Galton played a leading part in the dramatic capsizing of British liberalism's attitude to empire in the middle decades of the nineteenth century. Between 1848 and 1866, the 'progressive superintendence' of the Mills' humanitarian liberalism – which, as Mehta has argued, persuaded English reformers that they were towing 'societies stalled in their past in to contemporary time and history'[33] – was overthrown by a pessimistic and brutal justification of imperial domination as biologically ordained racial supremacy. Many forces contributed to this change. By the early 1850s key intellectuals – like Carlyle, Dickens, Froude (and Galton) – looked back at the optimism of abolitionism with inter-generational disdain. The catastrophic results of the grand plans for the 1841 Niger expedition marked the end of popular abolitionism and brought home the implications of disease for the British project of exploring and

---

[32] Whither Biometrics Committee, National Research Council, *Biometric Recognition: Challenges and Opportunities*, ed. Joseph N. Pato and Lynette I. Millett (Washington, DC: The National Academies Press, 2010).

[33] Uday Singh Mehta, *Liberalism and Empire: A Study in Nineteenth-Century British Liberal Thought* (University of Chicago Press, 1999), 81–2.

changing Africa. Meanwhile, at home the uprisings in France in the summer of 1848 reminded the propertied of the misery and dangers of working-class life in Britain, prompting Carlyle, in particular, to roar about the hypocrisy of the humanitarians worrying about the descendants of slaves in the colonies while the poor in Birmingham were starving in the gutter. But it was the Indian Revolt of 1857 that finally upended the humanitarians' argument of a carefully circumscribed rights-based empire, leaving in its place a lurid enthusiasm for the claim that might was right in the preservation of empire. By 1866 this argument was being fought out in England in the humanitarians' unsuccessful prosecution of Governor Eyre's brutal suppression of revolt in Jamaica. In this struggle it was Carlyle, and his disciples like Ruskin, who created the arguments and the popular interest in the hard-edged racial imperialism of Disraeli in the 1870s and Chamberlain in the 1890s.[34]

After the 1860s Social Darwinism gave scientific authority to the argument that the dominion of the fittest in the empire was sanctioned by human evolution. This outcome was horribly paradoxical, as Desmond and Moore have recently shown, for Darwin's life-project was motivated by the humanitarian imperative to demonstrate the common humanity of slaves and slave-owners. From his youth with the Wedgewoods in Clapham to his public denunciation of Eyre in 1866, Darwin was motivated by an intense personal horror of the brutality of slavery to disprove the phrenologists and polygenists arguing for its preservation.[35] In this Darwin's science – like John Herschel's – was crucially shaped by the experience of life in the colonies.[36] Yet, certainly by the time Darwin published *Descent of Man*, in 1871, Galton's eugenic arguments had persuaded him that 'the most able should not be prevented by laws or customs from succeeding best and rearing the largest number of offspring'.[37]

For Galton was also changed by his time in South Africa, and he would use that experience to redirect the politics of Darwin's science. Interleaved

[34] Bernard Semmel, *Jamaican Blood and Victorian Conscience: The Governor Eyre Controversy* (New York: Houghton Mifflin, 1963); Philip D. Curtin, *The Image of Africa: British Ideas and Action, 1780–1850*, 2 vols (Madison: University of Wisconsin Press, 1964); Catherine Hall, *Civilising Subjects: Metropole and Colony in the English Imagination 1830–1867* (Cambridge: Polity Press, 2002); Theodore Koditschek, *Liberalism, Imperialism, and the Historical Imagination: Nineteenth-Century Visions of a Greater Britain* (Cambridge University Press, 2011), 155–77.

[35] Desmond and Moore, *Darwin's Sacred Cause*.

[36] Musselman argues that Herschel adopted his distinctive form of inductive reasoning because of his growing distaste for the brutality of the frontier in South Africa, and, especially, its emphasis on hunting. Musselman, 'Swords into Ploughshares'.

[37] Charles Darwin, *The Descent of Man, and Selection in Relation to Sex* (London: D. Appleton, 1872), 461; Desmond and Moore, *Darwin's Sacred Cause*, 303.

with his pessimistic statistical explanation of the dangers of undirected human reproduction, he produced a deeply influential account of African inferiority and the pivotal claims of Social Darwinism. Galton's writing also became an authoritative ethnographic source on the moral and cultural features of the Herero people in particular; a fact that bears special weight in light of the German army's genocidal attack on them in 1904. It was also Galton who produced the key argument that races were defined by national populations formed by descent which, critically, would tend always to revert back to their normal characteristics formed over time.[38] And it was this claim that was used by the segregationists of Chamberlain's empire – half a century after Galton arrived in Cape Town – to set in place the biometric registration systems that created the distinctive form of the South African state.

Galton spent two years in South Africa: he left London on 5 April 1850 and arrived back in the city on the same day in 1852. For most of that time – August 1850 to January 1852 – he travelled in the interior of what is now Namibia, a 2,000-mile journey mostly on the back of an ox. Taking careful readings with a large sextant he carried with him, his expedition took him from Walvis Bay up to Ovamboland, very close to the border with Angola, and then across to the east into the Kalahari Desert. In the book he produced soon after returning to London, he described his objective as filling 'up that blank in our maps which, lying between the Cape Colony and the western Portuguese settlements, extends to the interior as far as the newly discovered Lake Ngami'.

Galton's *Narrative of an Explorer in Tropical South Africa* was published in 1853. The 300-page memoir is a detailed and engagingly written traveller's account, shaped by his amateur interests in hunting, geography and linguistics. The book became, as Stocking notes, one of the key sources of data for the armchair ethnologists of the 1860s.[39] But there is no sign in this work of Galton's later obsession with anthropology: 'not a word as to how to observe and record the anthropometric characters, folk-lore or religious customs of savage man; neither callipers, tape nor colour standards appear in Galton's instrumentarium.'[40] Nor is it particularly interesting as ethnography – Galton was much too preoccupied with the basic categories and mechanics of race to spend much time on the individuals he was encountering.

[38] Szreter, *Fertility, Class and Gender in Britain*, 101–18.
[39] Stocking, *Victorian Anthropology*, 80.
[40] Pearson, *The Life, Letters and Labours of Francis Galton: Researches of Middle Life*, vol. 2, 3–4. The often cited exception is his jocular use of a sextant to measure the anatomy of one of the local women. S.F. Galton, *The Narrative of an Explorer in Tropical South Africa* (London: J. Murray, 1853), 110.

Galton was accompanied on his expedition by the Anglo-Swedish adventurer, Charles Andersson, who published his own account. The distinctive absence of anything resembling ethnographic sympathy in Galton's book (and in the man himself) is very obvious after a reading of his companion's description of the same journey.[41] There was much about the lives of the Africans that Andersson found revolting – particularly the Herero practices of coating their bodies with ochre and abandoning the chronically ill to the wilderness, but his account is shaped by real sympathy and politeness that is completely lacking in Galton's book. The individuals and villages Andersson encountered were distinctive, with names, subtle allegiances, real sources of wealth and complex ritual and spiritual customs. Where Galton saw nothing but poverty and violence outside of the grain fields of Ovamboland, Andersson discussed at length the vast herds of cattle held by the dominant Herero kings and described, with simple pleasure, the food and drink prepared for them. On two separate occasions he comments on the sexual attractiveness of the women they encountered; Galton's comments on the same subject were framed by visceral and supercilious disgust.[42]

What Galton's book does very well is to provide a detailed first-hand account of the violence of the South African frontier. When he arrived at Walvis Bay on what is now the Namibian coast in 1850 Galton stepped into the front line of the expanding boundary of conquest and trade emanating from the Cape. The thirty-year epoch after 1820 was an especially horrible time in Southern Africa, and historians have debated whether the sources of the conflict should be attributed to the rise of the Zulu state or the expansion of trading, cattle-raiding and slavery as the emigrants from the Cape – white and brown – moved in to the interior.[43] Galton was certainly very aware that 'the country here is in the wildest disorder' and convinced that the protagonists of the violence were 'a set of lawless ruffians many of whose leaders were born in the Cape Colony' but that did little to temper his interest in presenting the region as a laboratory for racial conflict in the most sweeping terms.[44]

---

[41] Charles John Andersson, *Lake Ngami: Or, Explorations and Discoveries during Four Years' Wanderings in the Wilds of Southwestern Africa* (London: Dix, Edwards & Co., 1857).

[42] *Ibid.*, 195.

[43] For an excellent synthesis see Norman Etherington, *The Great Treks: The Transformation of Southern Africa, 1815–1854* (London: Longman, 2001); and for examples of the revisionist research Carolyn Hamilton, ed., *The Mfecane Aftermath: Reconstructive Debates in Southern African History* (Johannesburg: Witwatersrand University Press, 1995); and Elizabeth Eldredge and Fred Morton, eds, *Slavery in South Africa: Captive Labor on the Dutch Frontier* (Pietermaritzburg: University of Natal Press, 1994).

[44] Galton, *An Explorer in Tropical South Africa*, 89–90; Karl Pearson, *The Life, Letters and Labours of Francis Galton: Birth 1822 to Marriage 1853*, vol. 1 (Cambridge University Press, 1914), 225.

His letters home and his book describe the peoples of this region caught up in a conflict of Hobbesian dimensions. The primary villains in his story were the Oorlam commandos – the mixed descendants of Khoi and Nama – Christian, Dutch-speaking, horse-riding and gun-bearing, whom Galton saw as the 'offset' of the emigrant Boers.[45] Galton provided detailed descriptions of the black Boers' frenzied efforts to seize cattle from Ovaherero lineages, part of a massive shift in a regional economy that had been organised around pastoralism and long-distance trade, to pervasive cattle-raiding. This change in the basic pattern of accumulation was fuelled by the market for cattle in the Cape, which the Oorlams (and the other frontier raiders across the subcontinent) used to buy the guns and horses they needed to seize the enormous herds of the pastoralists.[46] Both groups of pastoralists, the black Boers and the Herero, fought episodically against another group – the Berg Dama – who spoke a Khoi-San language but lived in the mountains. And, moving fluidly between them all, was another group, called the Bushmen, who lived in or near to the deserts and without cattle. Far to the north he encountered the settled farmers of Ovamboland, who stayed clear of the cattle-raiding. It was these categories – Oorlam, Berg-Dama, Bushmen, Herero and Ovambo – that provided Galton with a racial hierarchy manifestly differentiated by culture and history.

Galton's first point, made repeatedly and with carefully considered detail, was that – as he explained to his mother towards the end of his trip – the black peoples of South Africa (with the partial exception of the Ovambo) were 'brutal and barbarous to an almost incredible degree'. He constructed the case for this ubiquitous brutality by scattering anecdotes of casual anatomical violence through the narrative of his journey. His story opens with a description of an Oorlam attack on a mission station that was their first destination. The attack took place while Galton and his party were still en route. He describes meeting two women on the road 'one with both legs cut off at her ankle and the other with one' which the Oorlams had cut 'off with their usual habit, in order to slip off the solid iron anklets that they wear'. He reports, as if he witnessed the incident,

[45] See Brigitte Lau, 'Conflict and Power in Nineteenth-Century Namibia', *The Journal of African History* 27, no. 1 (1986): 29–39, for a review of the peoples and politics of this period.

[46] Galton was witnessing the rapid depletion of what Wilmsen has called the 'surplus native product' of the desert fringe, a reservoir built up over generations. In his own grisly hunting operations around the waterholes of the western Kalahari he was an early participant in the destruction of another unrecognised subsistence reservoir. *Ibid.*, 30; Edwin N. Wilmsen, *Land Filled with Flies: A Political Economy of the Kalahari* (University of Chicago Press, 1989), 93–129; Galton, *An Explorer in Tropical South Africa*, 268–87.

that one of the Oorlam leader's sons – 'a hopeful youth' – walked up to an abandoned child and 'leisurely gouged out its eyes with a small stick'.[47]

Andersson's account of this event, like his description in general, is restrained. He condemns the brutality of the attack on the mission station, accusing them of indulging a 'savage thirst for blood' but his version has none of the graphic violence of Galton's account, and he provides a detailed explanation of the basis of the conflict between Jonker, commander of the Oorlams, and Kahichene, leader of the Ovaherero settlement at the mission. This is the real difference between the two descriptions. For Galton, events are driven by racial populations with undifferentiated qualities and a universal propensity for horrifying violence. The subjects of Andersson's book are individuals, with distinctive moral qualities. Jonker Afrikaner emerges as a ferocious, bloodthirsty, but very successful, individual leader of the raiders, quite unlike the other Oorlam leaders. There was no intrinsic tolerance for violence, and no racial solidarity. 'Jonker [Afrikaner] and [William] Zwartbooi', he explains, 'associated occasionally, but they were by no means well-disposed towards each other'.[48]

The historiography of nineteenth-century South Africa is replete with horrible violence,[49] but Galton's account is distinguished by the vicious sentiment he attributes to victims and perpetrators alike. In the 1870s Galton defended the Royal Geographical Society's feeble efforts to rescue David Livingstone – who had gone missing while looking for the source of the Nile in Central Africa – by publicly criticising Henry Stanley's famously successful effort as sensationalist.[50] But his own writing from this earlier period was – especially in comparison with other accounts from the same period – unmistakably sensational. Galton offers an uncensored eyewitness account of the state of nature. Early in the book he describes rescuing two men who had been attacked by their neighbours. 'The first man's throat was cut through', he reports in careful anatomical detail, and of the second, 'all the back sinews of his neck were severed to the bone, and the cut went round his neck, but only skin deep near the jugular vein and the wind pipe'. Galton has two points in mind here. The first was the claim, often repeated, that Africans were physiologically much stronger than Europeans: 'The tenacity of life in a

[47] Galton, *An Explorer in Tropical South Africa*, 66, 67.
[48] Andersson, *Lake Ngami*, 103.
[49] The bibliography here is potentially very long. The finest single study is Jeff B. Peires, *The Dead Will Arise: Nongqawuse and the Great Xhosa Cattle-Killing Movement of 1856–7* (Johannesburg: Ravan Press, 1989); see also Etherington, *The Great Treks*; and Eldredge and Morton, *Slavery in South Africa*.
[50] Gillham, *A Life of Sir Francis Galton*, 128, 131.

negro is wonderful.'[51] The second was that the black people he encountered lived in a world of Hobbesian brutality.

What distinguishes Galton's account from Andersson's is the vicious-
ness he attributes to the Herero pastoralists whose herds were being
plundered by the Nama raiders. Galton's book presents the life of the
pastoralists as a struggle not worth living. He describes in detail the
predicament of individuals attacked by hyaenas in their sleep, burned
almost to death by lightning strikes and abandoned by their people. In a
letter to his mother, written after he had been in the country for a year,
he presented the pastoralists as the agents of horrifying violence. 'The
Ovahereros, a very extended nation, attacked a village the other day for
fun, and after killing all the men and women', he wrote, 'they tied the
children's legs together by the ankles, and strung them head downwards
on a long pole, which they set horizontally between two trees; then they
got plenty of reeds together and put them underneath and lighted them;
and as the children were dying poor wretches, half burnt, half suffocated,
they danced and sung round them, and made a fine joke of it'.[52] Like
many of the events Galton mentions, the sources of his information are
obscure. But what is clear is that he viewed the black people of the desert
as morally depraved and incapable of sympathy. Galton returns, repeatedly, to the claim that 'the Damaras kill useless and worn-out people:
even sons smother their sick fathers'.[53]

Sensationalism about Africans, as Curtin observed a long time ago,
was a necessary ingredient of European travellers' memoirs well before
the nineteenth century, part of the formula of brutality and violence that
was required to sell to a popular market.[54] Yet Galton's account carries
more than the normal moral burden of a lurid traveller's tale because it
became an authoritative description of the Herero people, a key part of
what Steinmetz has called the *Devil's Handwriting*: the inter-generational
ethnographic literature on the Herero that was 'overwhelmingly hateful,
even exterminationist' decades before the events of 1904. The ethnog-
raphies were not responsible for General von Trotha's decision to drive
tens of thousands of Herero – men, women and children – into the desert,
but they did, as Steinmetz shows, remove the grounds for a humanitarian
defence of their right to live. Steinmetz does not stress Galton's place in

[51] This story is told in Pearson, *The Life, Letters and Labours of Francis Galton: Birth 1822
to Marriage 1853*, vol. 1, 30; and Galton, *An Explorer in Tropical South Africa*, 90–2, the
claim is repeated on 66, 224; and Galton, *Memories of My Life*, 36.
[52] Pearson, *The Life, Letters and Labours of Francis Galton: Birth 1822 to Marriage 1853*, vol.
1, 236.
[53] Galton, *An Explorer in Tropical South Africa*, 112–13, 190–1.
[54] Curtin, *Image of Africa*, 323.

the elaboration of the *Devil's Handwriting*, but the links between his narrative and the accounts of the 1870s and later were direct.[55]

But these appalling effects, if they are partially Galton's responsibility, were only felt in the last years of his life – the immediate lesson of his time in South Africa was more abstract: Galton saw in the racial tapestry of the frontier the evidence of the dialectic between biology and culture that became his obsession after 1865. He returned, repeatedly, to the self-evident instability of his own racial classifications.[56] On the one hand this was an argument about biological determinism. 'There is no difference between the Hottentot and the Bushman, who lives wild about the hills in this part of Africa, whatever may be said or written on the subject', he insisted:

The Namaqua Hottentot is simply the reclaimed and somewhat civilised Bushman, just as the Oerlams represent the same raw material under a slightly higher degree of polish. Not only are they identical in *features* and language but the Hottenot tribes have been, and continue to be, recruited from the Bushman ... In fact, a savage loses his name, 'Saen', which is the Hottentot word, as soon as he leaves his Bushman life and joins one of the larger tribes.

But on the other it was about the cultural and, in Galton's terms, psychological change. Thus he described the 'Namaqua "Oorlams" or Namaquas' as people 'born in or near the colony, often having Dutch blood and a good deal of Dutch character in their veins'.[57]

These changes in culture and character, from slave to master, were also unmistakably about the making of hierarchy. Galton insisted that the terms Oorlam, Hottentot and Bushman referred to 'the identically same yellow, flat-nosed, woolly-haired, clicking individual'. But what was at work was a scale of civilisation: 'the very highest point of the scale being a creature who has means of dressing himself respectably on Sundays and gala-days, and who knows something of reading and writing; the lowest point, a regular savage.'[58] Invoking the ranking logic that would later distinguish Galton's statistics he observed that 'all things are relative' and reported that 'what these Oerlams were to the Dutchmen, that the Namaqua Hottentots are to the Oerlams'.[59] Galton's ethnography is unmistakably about race, and racial order, but, unlike the other biological

[55] Steinmetz, *The Devil's Handwriting*, 1–239. Steinmetz does describe, without linking it to Galton, how Eugen Fischer, one of the leading Nazi eugenicists, learned his trade in the wreckage of the Herero genocide, see 235.

[56] For more examples of this obsession with the contradictory standing of biology and culture see Galton, *An Explorer in Tropical South Africa*, 49, 88, 117, 179, 230–2, 250; and Galton, *Memories of My Life*, 143–4.

[57] Galton, *An Explorer in Tropical South Africa*, 67–9.

[58] *Ibid.*, 69.     [59] *Ibid.*, 68.

racists of his time, he was absorbed by the contingency and instability of almost all the racial categories he encountered on the frontier. In the middle of the 1860s he was using these colonial examples to support the argument that 'the improvement of the breed of mankind is no insuperable difficulty'.[60]

Thirty years later, he would return to the same themes in *Inquiries into Human Faculty and its Development*, the book that introduced the word eugenics into the English language. As was his habit, Galton concluded the argument of the book – which included his suggestions for a national registry for improving the human stock of Britain and a frightening account of the biological causes of criminality that was illustrated using the misfortunes of the New York Jukes[61] – by turning to the lessons of his South African data. First, here, were the statistical implications of domesticated oxen (of which more below). But he then turned to an extended discussion of the population lessons from the empire. Galton's concluding chapter painted the same demographic landscape that he had described in his travel narrative:

Damara Land was inhabited by pastoral tribes of the brown Bantu race who were in continual war with various alternations of fortune, and the several tribes had special characteristics that were readily appreciated by themselves. On the tops of the escarped hills lived a fugitive black people speaking a vile dialect of Hottentot, and families of yellow Bushmen were found in the lowlands wherever the country was unsuited for the pastoral Damaras. Lastly, the steadily encroaching Namaquas, a superior Hottentot race, lived on the edge of the district. They had very much more civilisation than the Bushmen, and more than the Damaras, and they contained a large infusion of Dutch blood.

And he put the history of the Cape to use as a laboratory for eugenic science. The settlement at the Cape, he argued, 'is barely six generations old', but the evidence of a 'curious and continuous series of changes' in a struggle of the fittest amongst the 'strange medley of Hottentot, Bantu, Malay, and Negro elements' showed the power of social and cultural interventions over biology. The story of the Cape frontier showed 'that men of former generations have exercised enormous influence over the human stock of the present day, and that the average humanity of the world now and in future years is and will be very different to what it would have been if the action of our forefathers had been different'. This power over the 'future human stock' placed a 'great responsibility in the hands of each generation' to act.[62]

---

[60] Francis Galton, 'Hereditary Talent and Character', *Macmillan's Magazine*, 1865.

[61] Kevles, *In the Name of Eugenics*, 71.

[62] Francis Galton, *Inquiries Into Human Faculty and Its Development* (London: Macmillan, 1883), 206–7.

### Slavery and flogging

In stark contrast with the humanitarian motivations of Darwin's work, Galton – following Carlyle's infamous essay from a year earlier – was building the case for the defence of slavery, or, to be more exact, for the imposition of very coercive forms of labour mobilisation and control. This enthusiasm for slavery was partly youthful rebellion, a contumacious and self-conscious renunciation of the universalising morality of his Quaker ancestry.[63] But the arguments remained with him throughout his life and they had wide significance, in Britain and in the empire.

In direct opposition to those arguing at this time for a cross-cultural recognition of equality with Africans – like the non-conformist missionaries and the Anglican Bishop Colenso[64] – Galton's ethnographic account of the peoples of Hereroland presented a tendentious case that slavery was inescapable. He made this argument with several interwoven claims about African society. The first of these was that slavery was endemic. 'It is not easy to draw a line between slavery and servitude', but his experience led him to say that 'the relation of the master to the man was, at least in Damara and Hottentot land, that of owner rather than employer'.[65] Like the missionaries he attributed part of the pervasiveness of slavery to the frontier raiders. 'The Namaqua Hottentots and Oerlams, in all their plundering excursions, capture and drive back with them such Damara youths as they take a fancy to, and they keep them, and assert every right over them.' There was little controversial about that claim – it was a standard complaint directed at Trekkers across the subcontinent. But the next part of Galton's argument was certainly unprecedented. He alleged that individuals volunteered for subjection amongst the Oorlams, and that the same kinds of relationships existed between rich and poor amongst the Damaras. 'These savages', he italicised, 'court slavery'.[66]

---

[63] In 1840, at the apex of the abolitionist moment in Britain, Galton visited the slave markets at Istanbul and commented to his travelling companion that he wished he had brought an extra £50 to purchase a Circassian slave woman. After he arrived in Cape Town a decade later and had equipped himself with local servants, including an ex-slave from Mozambique, he wrote to his brother. 'I have a Black to look after my nine mules and horses', he boasted, 'He calls me "Massa" and that also is very pleasant'. Pearson, *The Life, Letters and Labours of Francis Galton: Birth 1822 to Marriage 1853*, vol. 1, 223.

[64] Jean Comaroff and John L. Comaroff, *Of Revelation and Revolution: Christianity, Colonialism, and Consciousness in South Africa*, vol. 2 (Chicago University Press, 1997), 198–252; Jeff Guy, *The Heretic: A Study of the Life of John William Colenso* (Johannesburg: Ravan Press, 1983), esp. 69–82; William M. Macmillan, *The Cape Colour Question: A Historical Survey* (New York: Humanities Press, 1927, reprinted 1968), 210–89.

[65] Galton, *An Explorer in Tropical South Africa*, 237.    [66] *Ibid.*, 232.

This condition of self-imposed bondage, he claimed, was pervasive across the continent. Perhaps invoking the example of West African pawning, he claimed that 'all over Africa one hears of "giving" men away'. These people had 'abandoned the trouble of thinking what ... to do from day to day'. In a neat codicil to Carlyle's argument that Africans had been created to be ruled by Europeans, Galton claimed that across the continent people routinely surrendered responsibility for governing themselves. 'The weight of independence is heavier than he likes, and he will not bear it', he waxed; 'he feels unsupported and lost as if alone in the world, and absolutely requires somebody to direct him'.[67]

Studiously ignoring the effects of frontier-generated violence on the prospects of meaningful independence, Galton reported that the Herero 'seem to be made for slavery, and naturally fall in to its ways' and that the Berg Damara were 'abused and tyrannised over by everybody, but servitude has become their nature'.[68] These observations were only reinforced by the structural political relationships he witnessed between the settled grain-growing Ovambo in the far north and the nomadic peoples of the desert. For both the travellers what made the Ovambo 'a very different style of natives from those with whom we had been accustomed' was their economic autonomy. Andersson was much more conscious of the fact that this 'determination and independence' was a product of Ovambo isolation from the raiders. For Galton these elements of character were racial – products of personality that were selected inter-generationally. Where Galton interpreted the relationships between the pastoralists and the grain-farmers as evidence of a racial hierarchy and endemic slavery, Andersson stressed the formal equality he found amongst the Ovambo. 'They treated all men equally well', he wrote, 'and even the so much-despised Hottentots ate out of the same dish and smoked out of the same pipe as themselves'.[69] It was Galton's account that became authoritative – seeping in to the enormously influential writings of his friends Spencer and Darwin.

Another important ideological product of Galton's book was his rehabilitation of the kind of imperial mastery that was expressed in the use of the whip. Flogging, and the systematic use of the whip as discipline, was anathema for the humanitarians, as Desmond and Moore note in their biography of Darwin.[70] For Galton it was a matter-of-fact

---

[67] *Ibid.*, 239.    [68] *Ibid.*, 239 and 257.
[69] Andersson, *Lake Ngami*, 139.
[70] This theme runs thick through Darwin's biography, see Desmond and Moore, *Darwin's Sacred Cause*, 15, 69, 118, 133, 314.

necessity of imperial government, an indispensable and celebrated accessory of the overwrought masculinity demanded by the imperial frontier. (Indeed it is no exaggeration to suggest that he was the author in this book and the next, entitled *The Art of Travel*,[71] of the aesthetic McClintock has described as *imperial leather*.[72]) His journey was punctuated by floggings delivered with varying degrees of brutality. These anecdotes range from the threatened whipping of impertinent guides to the event in Ovamboland that may have been responsible for the atmosphere of mistrust that both Galton and Andersson described. After they had been waiting for several days for an audience with Nangoro, Galton started to worry that the herdsman provided by his hosts was not pasturing his oxen properly; to alter the man's routine he 'took active measures upon his back and shoulders, to an extent that astonished the Ovampo and reformed the man'.[73] The entire trip was concluded with the 'business-like application of a new rhinoceros-hide whip' after Galton participated in an Oorlam assault on the homestead of a Ovaherero man who had killed Galton's favourite, but abandoned, oxen.[74]

When Galton published his most popular book in the middle 1850s, a guide to wandering on the fringes of the empire called *The Art of Travel*, he assured his countrymen that 'the system of life among savages' was a slightly romanticised version of Hobbes' state of nature. Quoting the same verses from Wordsworth's poem 'Rob Roy's Grave' that appear on the opening page of Walter Scott's romance of the Scottish outlaw, he explained that the rule of action for the English traveller was that 'they should take who have the power, And they should keep, who can'.[75] It is clear from the tone of his descriptions that these beatings upheld a particular kind of heroic masculinity that celebrated the physical strength of the master. But there were some obvious limits. Towards the end of the travel account he confronts a thief 'six foot five inches high and large in proportion' and 'dared not whip him'.[76] Similarly he complained, a little ironically, that 'I often wanted to punish the ladies of my party, and yet I could not make their husbands whip them for me, and of course I was far too gallant to have it done by any other hands'.[77] In his autobiography, written nearly sixty years after his travels in South Africa, Galton

[71] Francis Galton, *The Art of Travel; or Shifts and Contrivances Available in Wild Countries*, 1st edn (London: John Murray, 1855).

[72] On the whip see Anne McClintock, *Imperial Leather* (London: Routledge, 1995), 80.

[73] Galton, *An Explorer in Tropical South Africa*, 215, see also 157, 200, 241, 289.

[74] *Ibid.*, 289.      [75] Galton, *The Art of Travel*, 60.

[76] Galton, *An Explorer in Tropical South Africa*, 240–1.

[77] *Ibid.*, 199. For an overview of the literature on imperial masculinity, see Angela Woollacott, *Gender and Empire* (London: Palgrave Macmillan, 2006), 58–79.

described holding 'a little court of justice on most days, usually followed by corporal punishment, deftly administered'.[78]

Galton's endorsement of flogging – like Carlyle's almost simultaneous recommendation of the use of the plantation owners' 'beneficient whip' on the freed slaves of Jamaica – was a self-conscious rejection of the political and philosophical arguments that had been made by the abolitionists and the utilitarians.[79] This studied attachment to the whip on the imperial frontier followed decades of very public controversy about the place of flogging in the British military, and vigorous abolitionist agitation in his own home town about the cruelty of whipping.[80] For Galton Africans lived in a world which required an entirely new set of rules, where there was little time for the utilitarians' justification of government as progress or the humanitarians' defence of a divinely sanctioned and common humanity. These were ideas that he carried back with him to England after 1852, and they remained with him for half a century.

## Africans and the Athenaeum

An immediate result of Galton's South African journey was his transformation from an upper-middle-class delinquent into a member of the English scientific elite. In the year it took him to write up his book Galton offered two papers to the Royal Geographical Society and published a summary article of his journey, including a very detailed map and table of coordinates of the triangle of territory between Walvis Bay, Ovamboland and the western Kalahari. Before his book was even out Galton was awarded the society's highest honour – the Founders' Medal – 'for having, at his own cost and in furtherance of the expressed desire of this Society, fitted out an expedition to explore the interior of Southern Africa'. The award was given more for Galton's geographical than his anthropological achievements, for a journey 'upwards of 2000 miles as to enable the Royal Geographical Society to publish a valuable memoir and map' about 'a country hitherto unknown'.[81]

---

[78] Galton, *Memories of My Life*, 145.
[79] Thomas Carlyle, 'Occasional Discourse on the Negro Question', *Fraser's Magazine for Town and Country*, February 1849, 534, http://homepage.newschool.edu/het//texts/carlyle/carlodnq.htm; and for a synthesis of Carlyle's role in this political debate, Catherine Hall, 'The Economy of Intellectual Prestige: Thomas Carlyle, John Stuart Mill, and the Case of Governor Eyre', *Cultural Critique* no. 12 (Spring 1989): 167–96.
[80] J.R. Dinwiddy, 'The Early Nineteenth-Century Campaign against Flogging in the Army', *The English Historical Review* 97, no. 383 (April 1982): 308–31; Hall, *Civilising Subjects*, 111.
[81] 'Presentation of the Gold Medals', *Journal of the Royal Geographical Society of London* 23 (1853): lviii–lxi.

As Galton acknowledged in his autobiography it was the Royal Geographical Society medal that 'gave me an established position in the scientific world' and 'caused me to be elected a Fellow of the Royal Society in 1856'. Even more importantly the acclamation he received from the geographers resulted in his election to the Athenaeum Club – the jealously guarded social epicentre of the Victorian intellectual aristocracy. At the still callow age of thirty-four, Galton joined his half-cousin Charles Darwin, Thomas Carlyle, Herbert Spencer, Thomas Huxley and Disraeli at the heart of the rising English scientific and literary elite.[82] It was his rapid and early adoption by these institutions that gave Galton influence far beyond his scholarly achievements and nurtured his narcissistic obsession with the English scholarly establishment. After his death, Pearson observed that Galton had 'worshipped as one of simpler faith at his own peculiar shrine – a shrine dedicated to the genius of his race' – he could easily have located the altar in one of the lovely smoking rooms of the Athenaeum Club.[83]

When he published *Hereditary Genius* in 1869 Galton used the normal distribution, which he called the 'law of the deviation from an average', to make his point about the value of reputations across generations. Relying heavily on examination results from Cambridge and Sandhurst, which could easily be mapped against a normal curve, he argued that the distribution of 'natural gifts' followed the same pattern in society and over time. 'There must be', he suggested, 'a fairly constant average mental capacity in the inhabitants of the British Isles, and that the deviations from that average – upwards towards genius and downwards towards stupidity – must follow the law that governs deviations from all true averages'.[84] Using the scales A to G, and a to g, he produced a ranking order that mimicked the normal curve. He began with the 'four mediocre classes a, b, A, B', which combined included more than four-fifths of the total, drifting out to the 'truly eminent' in class F who were placed 'first in 4,000'. In an extensive list of celebrated lives, Galton found virtue in the evidence of nepotism, proving that a small number of families produced the bulk of acclaimed individuals through history.[85] He used this evidence to make the claim that eminent individuals in class G, and the even more exalted class X (all ranks above G), were descended from

[82] Galton, *Memories of My Life*, 153; Gillham, *A Life of Sir Francis Galton*, 105–6; Stocking, *Victorian Anthropology*, 96, 253.

[83] Pearson, *The Life, Letters and Labours of Francis Galton: Researches of Middle Life*, vol. 2, 98.

[84] S.F. Galton, *Hereditary Genius: An Inquiry into Its Laws and Consequences* (London: Macmillan, 1869), 32.

[85] *Ibid.*, 317.

eminent men and tended to produce eminent sons. (Women, in Galton's account, were theoretically bearers of ability but in practice they were simply vessels for the production of children.[86]) And he tried to show, with even less success, that 'with moderate care in preventing the more faulty members of the flock from breeding, so a race of gifted men might be obtained'.[87]

Galton had very little to say about the 'faulty members' of the English flock in this book, relying, instead, on a comparison between Africans and Athenians to set up his argument in favour of encouraging the fertility of the professions and placing the weak in 'celibate monasteries'.[88] The tool that Galton used for his comparison was 'the law of deviation from an average, to which I have already been much beholden'.[89] And history was, again, the source of his data. Africans undeniably had some outstanding leaders, like the Haitian revolutionaries, but Galton used these figures and an arbitrary calibration procedure to prove the dismal implications of the normal distribution. 'The negro race has occasionally, but very rarely, produced such men as Toussaint l'Ouverture, who are of our class F; that is to say, its X, or its total classes above G, appear to correspond with our F, showing a difference of not less than two grades between the black and white races, and it may be more.' He made much the same point about the evidence of Africans being 'good factors, thriving merchants, and otherwise considerably raised above the average of whites – that is to say, it can not unfrequently supply men corresponding to our class C, or even D'. But these men were 'classes E and F of the negro', proving that the 'average intellectual standard of the negro race is some two grades below our own'.[90]

Without anything resembling evidence for these claims, Galton once again used his South African experience to confirm his mathematical speculations. Following the logic of the normal curve on its negative axis, he claimed to speak with authority on the pervasiveness of stupidity. Without citing 'every book alluding to negro servants in America', he offered his own observations of the large proportion of Africans who were feeble-minded. 'I was myself much impressed by this fact during my travels in Africa', Galton explained: 'The mistakes the negroes made in their own matters, were so childish, stupid, and simpleton-like as frequently to make me ashamed of my own species, I do not think it any exaggeration to say, that their c is as low as our e, which would be a difference of two grades, as before.'[91]

---

[86] *Ibid.*, 63.     [87] *Ibid.*, 64.     [88] *Ibid.*, 362.
[89] *Ibid.*, 337.     [90] *Ibid.*, 340.     [91] *Ibid.*, 341.

Writing at a time when figures like Robert Knox and James Hunt were articulating a crude and violent form of racial supremacy (which Galton called the 'nonsensical sentiment of the present day') that threatened to overwhelm the gentle and cerebral tenor that was cultivated by the Darwinians at the Athenaeum Club – he was careful to leave the imperial implications of his argument to his readers.[92] He set the context of a world civilisation that was growing in power and scope, and he warned of the great global transformation that would bring. He positioned the English between the Africans and the Athenians, whose ability 'on the lowest possible estimate, [was] very nearly two grades higher than our own – that is, about as much as our race is above that of the African negro'. And he warned that the English, whose towns were 'crushing them into degeneracy', faced the same problem as the Athenians. The demographic lesson from antiquity was clear. If the 'high Athenian breed … had maintained its excellence, and had multiplied and spread over large countries, displacing inferior populations' it would have hastened the progress of human civilisation 'to a degree that transcends our powers of imagination'. It was the imperial project – which 'swept away in the short space of three centuries' the peoples of the 'North American Continent, in the West Indian Islands, in the Cape of Good Hope, in Australia, New Zealand, and Van Diemen's Land' – that required eugenic intervention to produce a class of leaders who could manage the 'complex affairs of the state at home and abroad' with ease.[93]

## Statistical revolution

The most important elements of Galton's eugenic political philosophy were already well formed by the middle of the 1860s.[94] These elements – an insistent biological determinism, a conviction that race implied that psychological characteristics like bravery and intelligence are inherited from long-running processes of descent, and, most importantly, the conviction that these characteristics could be manipulated in human populations by controlling reproduction – were all expressed in two papers published by *Macmillan's Magazine*, both written before 1865 but published later: 'Gregariousness in Cattle' and 'Hereditary Talent and Character'. The direct inspiration for the arguments of these articles was the publication of Darwin's *On the Origin of Species by Natural Selection*,

---

[92] *Ibid.*, 362; Stocking, *Victorian Anthropology*, 246–57.
[93] Galton, *Hereditary Genius*, 341–3.
[94] Galton, *Memories of My Life*, 293; Pearson, *The Life, Letters and Labours of Francis Galton: Researches of Middle Life*, vol. 2, 70.

which was released in 1859. Galton 'devoured its contents' and attributed his enthusiasm for their arguments to a 'bent of mind' that both men inherited from their grandfather.[95] But Galton's reversal of the logic of natural selection in human beings – arguing that mediocre human descent promised to eliminate the exceptionally fit representatives of the species – did not come from Darwin. That required another source.

It is possible to trace the influence of the South African frontier on Galton's development of the normal distribution as a tool for explaining Darwinian evolution by following his curious interest in the statistical lessons offered by the oxen of Damaraland. This gentlemanly agricultural enthusiasm for breeding livestock started long before he arrived in South Africa, and it is a persistent theme in *Tropical South Africa* and *Art of Travel*. In the early 1860s, like Darwin, he had begun to think about the research implications of the domestication of animals. These speculations were originally presented as a paper to the Ethnological Society in which he wondered what sociological evidence could be gathered from the different forms of animal domestication. He took up this theme with a very different line of analysis in a piece that was later published as 'Gregariousness in Cattle and in Men' in *Macmillan's Magazine* in 1871, and then again, a decade later, in his book *Inquiries into Human Faculty*. Picking up on the politics of ranking along the normal distribution that motivated *Hereditary Genius*, and unmistakably reflecting his own anxieties about the expansion of democracy and the implications of natural selection, he reversed the logic of the original paper and used the behaviour of animals to construct arguments about the biology of human society.

The target of this paper was the 'natural tendency of the vast majority of our race to shrink from the responsibility of standing and acting alone'. Here he worried, with Gladstone's stumbling movement towards male democracy in the background, why most people exalted the 'vox populi, even when they know it to be the utterance of a mob of nobodies'. For the answer he turned to his African experience, and to the behaviour of other animals he knew well. Having had 'only too much leisure to think about them' during his slow journeys, he declared that the 'ox of the wild parts of western South Africa' was the other gregarious creature 'into whose psychology I am conscious of having penetrated most thoroughly'.[96] Galton used his rich familiarity with the cattle of Damaraland to show that it was the 'herd instinct', as Karl Pearson noted approvingly,

---

[95] Galton, *Memories of My Life*, 293.
[96] Francis Galton, 'Gregariousness in Cattle and in Men', *Macmillan's Magazine*, 1871, 353.

that was the source of 'many of man's intellectual weaknesses' under the conditions of modern civilisation.

Galton used the cattle obsession of the South African frontier to present a biology matched to the Victorian intelligentsia's worries, after Carlyle, about the origins and prospects of heroic leadership.[97] Remembering the tedious difficulties he had suffered training teams of oxen to pull their wagons in Ovamboland, he explained that a 'good "fore-ox" is an animal of exceptional disposition; he is, in reality, a born leader of oxen'. He found the evidence for biologically determined personalities in the selection practices of the frontier waggoners who spent their days looking for potential *voorosse*, beasts 'who show a self-reliant nature by grazing apart from the rest'. Even more prized were oxen who would tolerate being saddled, and ridden away from their companions. These colonial travails gave him real numbers to apply his new preoccupation with the Darwinian implications of the normal distribution. 'Why is the range of deviation from the average such that we find about one ox out of fifty to possess sufficient independence of character to serve as a pretty good fore-ox?' he asked. Galton saw an ominous lesson in these statistics which showed that 'natural selection tends to give but one leader to each herd, and to repress superabundant leaders'.[98]

To make his point about an evolutionary bias towards 'slavishness' he moved quickly from his discussion of the cattle of Damaraland to an account of the 'inhabitants of the very same country'. In looking for evidence of a European past in the present organisation of African societies and in accounting for the character of metropolitan culture on the basis of the reported behaviour of people outside of Europe, Galton was following the strong current of contemporary evolutionism. His innovation, like Malthus' a half century before him, was to assert the dismal prospects of progress using the new logics of natural selection and the normal distribution.[99] Finding the same 'gregarious instincts' amongst the Africans, he suggested that the people of Damaraland provided a laboratory for the investigation of the 'clannish fighting habits of our forefathers'. These 'blind instincts' for the protection of other human beings, produced by generations of barbarism, had the effect of 'destroying the self-reliant, and therefore the nobler races of men'. Implicitly comparing the English with the miserable conditions of the people of the desert, he concluded that a 'really intelligent nation' needed to break free from the biological constraints of the gregarious instinct. In order for the

[97] Frank M. Turner, 'Victorian Scientific Naturalism and Thomas Carlyle', *Victorian Studies* 18, no. 3 (March 1975): 332.
[98] Galton, 'Gregariousness in Cattle and in Men', 353–5.
[99] Stocking, *Victorian Anthropology*, 144–72.

English to escape their fate as a 'mob of slaves, clinging together, incapable of self-government' the instinct to subordination would need to be bred out, and the 'most likely nest' of these new heroic natures would be in the colonies.[100]

This curious interest in the socio-biology of Namibian cattle is an important, and mostly forgotten, window into the workings of Galton's combined obsessions with the empire and the normal curve. It was not a tangential interest amidst the broader project of his anthropometric interest in the English elite or the mathematics of correlation. Perhaps because he had such difficulty generating anthropometric data that would work with his normal distribution, Galton returned to the lessons of the Hereroland cattle in the early 1880s, thirty years after he had returned from South Africa. *Inquiries into Human Faculty* has two extended discussions of the lessons of cattle-breeding in Damaraland. In the opening chapters he used the original *Macmillan's* essay unchanged, and then, again, in one of the longest chapters of the book, he returned to the domestication of animals. 'The tamest cattle – those that seldom ran away, that kept the flock together and led them homewards – would be preserved alive longer than any of the others', he observes in conclusion; 'It is therefore these that chiefly become the parents of stock, and bequeath their domestic aptitudes to the future herd'. And he asserted the special importance of cattle selection as a test of the effects of heredity. 'I have constantly witnessed this process of selection among the pastoral savages of South Africa', he asserted, shifting in to an ambiguous tense: 'I believe it to be a very important one, on account of its rigour and its regularity.'[101]

It was this bizarre argument about the sociological implications of the behaviour of oxen that Darwin adopted in *Descent of Man* in his effort to account for the destructive moral effects of the sympathetic instincts in animals and men. Drawing on the authority of 'Mr. Galton, who has had excellent opportunities for observing the half-wild cattle in S. Africa', he quoted from the 1871 paper's claims 'that they cannot endure even a momentary separation from the herd'. The partly domesticated cattle (like the uncivilised tribes that formed the focus of this chapter) 'are essentially slavish, and accept the common determination, seeking no better lot than to be led by any one ox who has enough self-reliance to accept the position'. Lions otherwise quickly dispatch any who might have the independence to wander from the herd. Drawing on both his humanitarianism and Galton's pessimism, Darwin concluded that what

[100]  Galton, 'Gregariousness in Cattle and in Men', 357.
[101]  Galton, *Inquiries Into Human Faculty and Its Development*, 50–7, 193.

distinguished the savage from the civilised was 'the confinement of sympathy to the same tribe' and 'reasoning insufficient' to recognise the benefits of the 'self-regarding virtues, on the general welfare'.[102] Working in combination with the idea – also attributed to Galton – that 'with savages, the weak in body or mind are soon eliminated; and those that survive commonly exhibit a vigorous state of health', Darwin worried that the host of nineteenth-century public health interventions meant that 'the weak members of civilised societies propagate their kind'.[103] A gentler, more humanitarian (and abolitionist) temperament remained in *Descent of Man* – pointing out that the Athenians may have expired precisely because of their dependence on slavery – but the imprint of Galton's imperial pessimism is obvious throughout.[104] It was this new eugenic pessimism – derived from first-hand evidence of the empire – that sustained and endorsed the Social-Darwinist Darwin, overthrowing the political trajectory of his entire life's work. The political effects of the Social Darwinist rejection of humanitarianism were felt powerfully around the world in the generation that followed, but nowhere more powerfully than in South Africa.[105]

## Normal racism

In their histories of the racist thought that emerged after Darwin, Curtin, Stocking, Stepan and Dubow have each pointed to the special role that the disbarred Scottish surgeon, Robert Knox, played in articulating the break with the argument that human beings had a common biological nature, that 'all men were originally one'.[106] Curtin described Knox as 'the real founder of British racism and one of the key figures in the general Western movement toward a dogmatic pseudo-scientific racism'. Galton and Knox took up very different public positions in the emerging field of British Anthropology – Knox was one of the leading lights of the 'Anthropologicals' whose recommendations on colonial policy can probably best be described as genocidal. Galton became secretary of the 'Ethnologicals', while advocating policies that Knox would have

---

[102] Darwin, *The Descent of Man*, 86, 97–8.
[103] *Ibid.*, 109. This idea of natural selection in African society became a cornerstone of twentieth-century eugenics in South Africa, Dubow, *Illicit Union*, 154, 167–8.
[104] Darwin, *The Descent of Man*, 109–14, 431, 461.
[105] M. Lake and H. Reynolds, *Drawing the Global Colour Line: White Men's Countries and the International Challenge of Racial Equality* (Cambridge University Press, 2008), 75–113; Dubow, *Illicit Union*, 129–39; Koditschek, *Liberalism, Imperialism*, 212–39.
[106] Curtin, *Image of Africa*, 372; Stocking, *Victorian Anthropology*, 64–5; Stepan, *The Idea of Race in Science*, 41–3; Dubow, *Illicit Union*, 13–63.

endorsed, he believed that the study of the empire's colonial subjects should begin with the idea of common humanity.[107]

Knox took to lecturing publicly (and to polygenesis – the argument that the different races of man were distinct species) only after he was disbarred in 1828 for buying cadavers from a gang of murderers that he took to be grave-robbers.[108] His seminal work, *The Races of Men: A Fragment*, published while Galton was still wandering the desert on his ox, drew on his experience of the comparative anatomy of the peoples of the South African frontier to make the argument that Galton would later adopt as his own, that 'race or hereditary descent is everything'. Knox used his personal knowledge of the peoples of the Cape – 'the Hottentot, the Bosjeman, the Amakoso Caffre and the Dutch Saxon' – to make the point that environmental change and racial intermarriage were violations of the laws of physiology and history. After lecturing his audience on their physiological distinctiveness, he asked: 'Whence came these Bosjemen and Hottentots? They differ as much from their fellow-men as the animals of Southern Africa do from those of South America.' The history of the country, as a racial laboratory, seemed to confirm his belief that race-mixing was a demographic dead-end. 'The Dutch families who settled in Southern Africa three hundred years ago', he reported, 'are now as fair, and as pure in Saxon blood, as the native Hollander.'[109]

Both men drew their interpretations of racial biology from the same frontier society and, particularly, from the disturbing position of the light-skinned Khoi-Khoi and San in Blumenbach's[110] five-term continental racial taxonomy. But they differed crucially on the implications that South African history offered for the mutability of biological descent. For Knox, a radically pessimistic conservative, there was no possibility of change – distinctive and specific races 'up to the earliest recorded time, did not differ materially'; for Galton the 'curious and continuous changes' of the brown people of the Cape showed that 'men of former generations have exercised enormous influence over the human stock of

---

[107] Stocking, *Victorian Anthropology*, 250–3.

[108] See the paper he presented in 1822, which struggles to fit the peoples of South Africa into Blumenbach's taxonomy. 'We may view the human race as derived originally from one stock, to which the arbitrary name of Caucasian has been given.' Robert Knox, 'Inquiry into the Origin and Characteristic Differences of the Native Races Inhabiting the Extra-Tropical Part of Southern Africa', *Memoirs of the Wernerian Natural History Society* V, no. 1 (1824): 210.

[109] Robert Knox, *The Races of Men: A Fragment* (London: H. Renshaw, 1850), 65–6.

[110] Stephen J. Gould, 'The Geometer of Race', *Discover* (November 1994): 65–9; Knox, 'Inquiry into the Origin and Characteristic Differences of the Native Races Inhabiting the Extra-Tropical Part of Southern Africa'; Galton, *An Explorer in Tropical South Africa*, 232, 250.

the present day'.[111] The difference, of course, was Darwinism – Galton only began to formulate his arguments about hereditary ability after the publication of *Origin of Species* in 1859. This delay created the space for Galton's eugenic fantasy, but in most other respects the two men shared a common imperial politics. (Knox's account of the AmaXhosa people of the Eastern Cape, with whom Britain had been at war for a generation, was considerably more sympathetic than Galton's description of the unfamiliar Ovaherero and Ovambo peoples.)[112]

Galton's African experience was important in the development of his biometric statistics, but it was also much more significant in the development of racism than scholars have yet recognised. In all of his important works, with the exception of *Natural Inheritance*, Galton relied heavily on the evidence of his travels in South Africa to build his case for eugenic reform. The operations of the normal curve, decades before he had worked out the mathematics, were established using claims based on his African travels. More importantly, Galton inserted an entirely new conceptual weapon into the politics of race. The hereditary, and statistical, concept that Galton first developed in his discussion of regression, with its flexible present but ineluctable biological centre of gravity in the past and the future, broke with the older physiological, linguistic and geographical definitions of race.

It is important to notice that this argument about the normal distribution of racial qualities began very early in Galton's work. He started building the argument using his ethnographic experience decades before he had demonstrated the statistics of regression from the breeding of sweet peas. Early signs appeared in his response to the announcement of James Hunt's brutal polygenic racism in 1863.[113] For Hunt, like Edward Long, Samuel Morton and Robert Knox, physiology placed hard limits on the intellectual and cultural capacity of Africans. At a session of the British Association in Newcastle, which was 'interrupted by hisses and counter-cheers', Galton insisted that the humanitarians' demand that the 'negro only requires an opportunity for becoming civilized' was delusionary, and concluded that it was one of the 'decrees of Nature's laws' that the 'European [was] the conqueror and the dominant race'. But it

[111] Stocking, *Victorian Anthropology*, 65; Knox, *The Races of Men*, 70; Galton, *Inquiries Into Human Faculty and Its Development*, 208, 210.

[112] Knox on the AmaXhosa: 'They are circumcised, eat no fish nor fowl, nor unclean beasts, as they are called; live much on milk, and seem to me capable of being educated and partly civilized ... Their language is soft and melodious, and they seem to have an ear for simple melody ... by coming into contact with Europeans, they have become treacherous, bloody, and thoroughly savage. Yet they have great and good points about them.' Knox, *The Races of Men*, 159–60.

[113] On Hunt's views see Stocking, *Victorian Anthropology*, 249–51.

also entailed another, easily contradicted, assertion. 'The many cases of civilized blacks are not pure negroes', he announced, 'but, in nearly every case where they had become men of mark, they had European blood in their veins'. Galton's response suggests that he was already formulating his views on the normal distribution of human abilities. On the basis of his personal experience he claimed that Africa was home to 'more abject, superstitious, and brutal tribes than elsewhere in the world'. He was careful to avoid the obvious pitfall of Hunt's claims about African ability, arguing that he 'thought that occasionally the race had produced clever men' but not to a degree that would mitigate the 'slavish and brutal condition of the vast majority of the African race'.[114]

It was his standing as a moderate, cerebral, opponent of the crude arguments of the physical Anthropologists that gave Galton his special status, made him an authority on Africa and Africans, and a key figure in the marriage of racism and evolutionist thought in the 1870s and 1880s.[115] This scholarly standing gave incontrovertible weight to the most tendentious and execrable claims about Africans in general and the Herero in particular, especially in Herbert Spencer's massive comparative anthropologies. In Spencer's 1873 *Study in Sociology*, written for a popular audience in the United States, Galton is not credited but he is certainly the source of the claim that 'among the South-African races, a white master who does not thrash his men, is ridiculed and reproached by them as not worthy to be called a master'.[116] And a few years later, in Volume 1 of the *Principles of Sociology*, Spencer's opus magnum, Galton is acknowledged as the source for a litany of damning moral assessments. The Herero 'have no independence', they 'court slavery', 'admiration and fear being their only strong sentiments'.[117] They were presented as the exemplars of a pervasive and degenerate brutality. 'Galton', Spencer reminded his readers, 'condemns the Damaras as "worthless, thieving, and murderous"'. Again the most outrageous claims from Galton's *Narrative* were presented as unexceptional facts. The Herero mind was incapable of abstraction and arithmetic when 'no spare hand remains to grasp and secure the fingers'. Galton (like the other modern explorers) was cited repeatedly and authoritatively as a sociological witness 'trained in science'.

---

[114] 'British Association at Newcastle – Sectional Reports Continued', *The Reader*, 19 September 1863.
[115] For the effects of which see Dubow, *Illicit Union*, 18.
[116] Herbert Spencer, *The Study of Sociology* (New York: D Appleton and Company, 1878), 122.
[117] Herbert Spencer, *Principles of Sociology:Volume 1* (New York: D Appleton and Company, 1912), 63, 69.

Herbert Spencer has so disappeared from our contemporary social theory that it is difficult to to credit how immensely influential he was in his own lifetime, especially in the United States, but also in Britain, and even in Meiji Japan. He was, as Beatrice Webb put it, 'part of the mental atmosphere' of the late nineteenth century.[118] Spencerian common sense was especially powerful amongst the advocates of the scientifically driven technological change of the nineteenth century – like Andrew Carnegie and the mining engineers who flocked to South Africa.[119] Spencer offered a reassuring explanation of the importance of individualism and the inevitability of progress that could be deployed to oppose utilitarian reforms. For the wealthy middle class the Spencerian struggle was already in the past.

But Spencerian arguments, especially in South Africa (and in the South) presented two sinister corollaries: the first was often expressed by the advocates of segregation. The American mining engineer, William Honnold – founder of the Anglo-American Corporation, and an authority in Johannesburg, after 1900, on black people on both sides of the Atlantic, built his case for coercive segregation on a synthesis of Galton and Spencer. For Africans, he argued, the 'struggle for existence had been along such simple lines, and after so spiritless a fashion that there was little chance for the correcting influence of natural selection'. Even more worrying, this languid order had existed for so long 'as to fix in the race a fundamental ineptitude for progress'.[120] And, second, as Mohandas K. Gandhi protested in his struggles against the Transvaal (see Chapter 3), the doctrine of the survival of the fittest provided a handy justification for very coercive remedies to the Spencerian problem of conflicting levels of civilisation.

As we will see in the next chapter, it was Francis Galton who first recommended the introduction of a system of compulsory fingerprinting in South Africa, relying on a policeman from Bengal to do it. Galton's most significant contribution to the evolution of the South African state lay in providing the liberal progressives, who arrived on the Witwatersrand from Oxford and Cambridge with Alfred Milner in 1900 with a concept

[118] Beatrice Potter Webb, *All the Good Things of Life, 1892–1905*, ed. Norman Ian MacKenzie and Jeanne MacKenzie (Cambridge, MA: Harvard University Press, 1983), 286.

[119] Edwin T. Layton, *The Revolt of the Engineers; Social Responsibility and the American Engineering Profession* (Cleveland: Press of Case Western Reserve University, 1971), 80–1; Clark C. Spence, *Mining Engineers & the American West; the Lace-Boot Brigade, 1849–1933* (New Haven: Yale University Press, 1970), 333–7; Elaine N. Katz, 'Revisiting the Origins of the Industrial Colour Bar in the Witwatersrand Gold Mining Industry, 1891–1899', *Journal of Southern African Studies* 25, no. 1 (1999): 73–97.

[120] William L. Honnold, 'The Negro in America', n.d., Fortnightly Club, Johannesburg, South African Historical Archive, University of the Witwatersrand.

of race that was purpose-built to the objects of the segregationist state. Chief amongst these progressives was Lionel Curtis. A disciple of Octavia Hill, the architect of the white working-class welfare state in the city of Johannesburg before 1903, Curtis would go on to become Gandhi's tormentor, and an architect of the constitution of the new Union of South Africa in 1910. After he left South Africa Curtis would become one of the stalwarts of Milner's Round Table movement, the Beit Lecturer in colonial history at Oxford, inventor of the mad scheme of dyarchy in India in the 1920s, and, in the 1940s, of world government in a (failed) imperial vein.[121]

It was Curtis who first used the Asiatic Registry – the compulsory ten-print, fingerprint register in South Africa – against Gandhi. He was also the most influential political theorist of racial segregation in Johannesburg before 1910.[122] Curtis made the case for the segregationist state, for abandoning the 'splendied theory of absolute equality for all British subjects' in a famous essay titled, ominously, 'The Place of the Subject Peoples in the Empire'.[123] In this essay Curtis presented the Galtonian argument of racial regression to justify using compulsory fingerprinting to 'shut the gate' of the segregationist state. Drawing on the newly discovered virtues of the Orange Free State as a self-governing republic peopled by 'Boers, few of whom could read or write', Curtis rejected the Macaulayite principle of extending rights to those who mastered the liberal tests of citizenship. In the midst of the struggle with Gandhi over the failed promise of humanitarian liberalism, like the opponents of the Bengali babus,[124] he rejected the idea that 'educational superiority' should imply citizenship. 'From the social and political point of view individuals must be judged not by what they are but by their potentiality and that potentiality can only be measured by the history of the race as a whole', he explained. 'An individual may rise far above the level of his race, but he cannot raise his posterity with him.'

Curtis then went on to lay out the blueprint of what would gradually become the Apartheid state. 'Given one authority in South Africa with

[121] D. Lavin, *From Empire to International Commonwealth: A Biography of Lionel Curtis* (New York: Oxford University Press, 1995); Lionel Curtis, *With Milner in South Africa*, n.d.

[122] Shula Marks, 'War and Union, 1899–1910', in *Cambridge History of South Africa, 1885–1994*, vol. 2 (Cambridge University Press, 2012), 166–8.

[123] Martin Legassick, 'British Hegemony and the Origins of Segregation in South Africa, 1901–1914', in *Segregation and Apartheid in Twentieth Century South Africa*, ed. William Beinart and Saul Dubow (London: Routledge, 1995), 43–59; Dubow, *Illicit Union*, 130.

[124] M. Sinha, *Colonial Masculinity: The 'Manly Englishman' and the 'Effeminate Bengali' in the Late Nineteenth Century* (Manchester: Manchester University Press, 1995); Koditschek, *Liberalism, Imperialism*, 320–5.

a strong hand guided for generations by a clear purpose there is nothing to prevent the gradual segregation of the natives in the great locations already set apart', he observed with chilling foresight. 'I am picturing a state of affairs in which the native is free to move about South Africa but has been led to fix his home in native territory and to find himself in the position of an *uitlander* when he goes outside it.'[125] Doing those two things – fixing the natives in the countryside and treating them as foreigners when they were in the cities – required a special kind of state, one which was already well under development in Johannesburg.[126]

[125] Lionel Curtis, 'The Place of Subject People in the Empire', 9 May 1907, 25–6, A146, Fortnightly Club, Johannesburg, South African Historical Archive.

[126] In the long run, Galton's eugenics had, at best, an ambiguous hold on South Africa. The supremacy of Mendelism after 1910 quickly supplanted Galton's simple biology in the most popular eugenic prescriptions. Dudley Kidd, *Kafir Socialism and the Dawn of Individualism; an Introduction to the Study of the Native Problem* (London: A. and C. Black, 1908). Two alternative views of race predominated. The first of these was Knox's brutal physical account of race fostered by anatomists trained in Edinburgh, a continuous stream of fossil discoveries, the growth of paleoanthropology as a field, and the increasingly mythical status of the San as 'living fossils'. The second was derived from history, and particularly from the German romantic obsession with tribal migration, which provided an essential justification for colonial settlement. Between them these models of race dominated South Africa after 1900. Dubow, *Illicit Union*, 18–129. Similarly, enthusiasm for eugenic explanations in the justifications for urban segregation, public health advocacy and some enthusiasts for intelligence testing, ran up against the insurmountable obstacle of enfranchised poor whites – the most obvious targets of compulsory sterilisation, an identical problem faced by those seeking to use racial descent for the definition of the franchise, deftly solved in the 1960s by tracing descent only to the 1951 census (see Chapter 6). *Ibid.*, 154.

## 2    Asiatic despotism: Edward Henry on the Witwatersrand

Francis Galton invented modern fingerprinting, but he did it using the workshop that colonial government provided in nineteenth-century India. This is a story that has been told so often, and so well, that it will only bear repeating in the most schematic form here.[1] Using a global network of courts and prisons for the enforcement of contracts for indentured labour, private debts, land revenues and long-term criminal sentences, the Indian colonial state created a global infrastructure of coercive identity registration. All of the accounts of the emergence of colonial fingerprinting stress the power of the logic of writing in shaping the trajectory of coercive systems of registration in India. From the use of tattoos to mark the faces of thugs and perjurers in the first half of the nineteenth century, the registration of entire populations of criminal tribes, through the use of photographs and the precise physiological measurements of Bertillonage generating distinguishing numerical indices, historians have traced a sequence of inscriptions and abstractions that eventually produced a working method of fingerprinting in Bengal in the 1890s.[2]

Yet, in their interest in the effects of writing on the organisation and abstraction of registration these scholars have tended to neglect an important part of Galton's proposals, one which has been sustained into the present: fingerprinting was designed as a remedy for the absence of writing. From his very first papers, Galton directed fingerprinting at the illiterate populations of the empire. 'It would be of continual good service in our tropical settlements, where the individual members of the swarms of dark and yellow-skinned races are mostly unable to sign their

[1] Radhika Singha, 'Settle, Mobilize, Verify: Identification Practices in Colonial India', *Studies in History* 16, no. 2 (2000): 151–98; Simon A. Cole, *Suspect Identities: A History of Fingerprinting and Criminal Identification* (Cambridge, MA: Harvard University Press, 2001), 60–96; Chandak Sengoopta, *Imprint of the Raj: How Fingerprinting Was Born in Colonial India* (London: Macmillan, 2003); Clare Anderson, *Legible Bodies: Race, Criminality, and Colonialism in South Asia* (Oxford and New York: Berg, 2004).

[2] In addition to those cited above, see Carlo Ginzburg, 'Clues: Roots of an Evidential Paradigm', in *Clues, Myths and the Historical Method* (Baltimore: Johns Hopkins University Press, 1989); Allan Sekula, 'The Body and the Archive', *October* 39 (1986): 3–64.

names and are otherwise hardly distinguishable by Europeans', he wrote in 1891, 'and, whether they can write or not, are grossly addicted to personation and other varieties of fraudulent practice'.[3]

The practical origins of fingerprinting lie in the harrowing struggles between landlords, tenants and the colonial state in Bengal in the years immediately after the 1857 Rebellion. It was there that Sir William Herschel, the oldest son of the Cape astronomer and Chief Magistrate of the Hooghly district, first resorted to 'taking the signature of the hand itself' in an effort to frighten a contractor providing materials for road building 'from afterward denying his formal act'.[4] Soon afterwards Herschel found himself enforcing the brutal contracts for indigo between Zemindars, peasants and manufacturers, a struggle that made written documents of rents and receipts 'worth no more than the paper on which they were written'.[5]

In the monographs and papers that Galton published between 1892 and 1895 – *Finger Prints*, a supplementary chapter called 'Decipherment of Blurred Finger-Prints', and *Finger Print Directories* – he drew on this administrative history to make the key point that fingerprints were unchanging. And it was these published works that prompted the widespread institutional enthusiasm for fingerprinting in the English colonial world, including back in Bengal where they were read by Edward Henry, the Commissioner of Police. Galton's scheme had many arresting features, but at its heart it was designed as a tool for strengthening the imperial bureaucracy.

To demonstrate the viability of fingerprints as a mechanism for establishing and fixing the identity of very large numbers of people, and to supplant alternative systems, like Bertillon's anthropometric photography, Galton needed to rid the identity of each fingerprint of all ambiguity and establish a mechanism for storing each fingerprint record that would allow for rapid and accurate recovery. He did this by focusing on the seam that cuts through the complex patterns on the finger, leaving in place a single dominant pattern. 'After a pattern has been treated in this way', Galton reassured his readers, 'there is no further occasion to pore minutely into the finger print, in order to classify it correctly'.[6] The single

---

[3] Francis Galton, 'Identification by Finger Tips', *Nineteenth Century* 30 (August 1891): 303–11.

[4] Francis Galton, *Finger Prints* (London and New York: Macmillan and Co., 1892), 27–8; discussed in Sengoopta, *Imprint of the Raj*, 54–78.

[5] William James Herschel, *The Origin of Finger-Printing* (London: Oxford University Press, 1916), 11, www.archive.org/details/originoffingerpr00hersrich; Tirthankar Roy, 'Indigo and Law in Colonial India', *The Economic History Review* 64 (2011): 60–75; Sengoopta, *Imprint of the Raj*, 62–77.

[6] Galton, *Finger Prints*, 69.

remaining pattern, skewered by the intersection of the three corners of the seam, was then classified as one of three archetypes: Arch, Loop or Whorl. While Galton's scheme allowed for an almost infinite elaboration of sub-classifications (for example, the Forked Arch, Eyeletted Loop, Ellipses Whorl) these three basic categories gave him a mechanism for converting each finger into a letter, A, L or W. 'The bold firm curves of the outline', he explained, 'are even more distinct than the largest capital letters in the title-page of a book'.[7] It was a simple matter, thereafter, to make each hand into a word that could be classified alphabetically.

Here lay the key to the unusual power of Galton's system: his finger-printing classification provided a simple mechanism for converting the obscure qualities of the body into a textual object, subject to the normal procedures of indexing that were being used widely by the documentary bureaucracy. (Bertillonage did something very similar but required a much more complex examination.) It was Galton who popularised the stunning claim that properly classified fingerprints will produce a mathematically unique identifier for every human being – what we today would call a unique biometric identifier. In theory, and in practice, Galton provided a means whereby 'a fingerprint may be so described by a few letters that it can be easily searched for and found in any large collection, just as the name of a person is found in a directory'.[8] Yet fingerprints, unlike names, were physically bound to the person they denominated, and free of the ambiguity and manipulation that characterised naming. They provided, as Galton put it, 'a sign manual that differentiates the person who made it, throughout the whole of his life from the rest of mankind'.[9]

But the fingerprint classification scheme that Galton publicised in the early 1890s did not work. In even relatively small populations the three-letter, ten-digit words his system used did not form unique labels, and in large populations (of tens of thousands of people) his classifications produced large groups of identical records. To sort these groups of duplicates he produced a further set of sub-classifications but they were so tricky to use that he could not agree even with his own assistant on their application. It was this complexity, and uncertainty, of classification that prompted the British Home Office in 1893 to retain the more straightforward and tedious measuring rituals of Bertillon's anthropometrics as the basis of criminal identification. And, despite the claims of his earlier book, in 1900 Galton himself was still anxious about abandoning the elaborate certainties of Bertillonage for the single-step solution of fingerprinting.[10]

---

[7] *Ibid.*  [8] *Ibid.*, 13.  [9] *Ibid.*, 4.
[10] Cole, *Suspect Identities*, 80.

Towards the beginning of the 1890s, Edward Henry, working as the Commissioner of Police in Bengal, began to implement Bertillon's system of identification and registration. 'With anthropometry on a sound basis', he reported, 'professional criminals of this type will cease to flourish, as under the rules all persons not identified must be measured, and reference concerning them made to the Central Bureau'.[11] Very quickly the Bengal police built up a register of the 'principal criminals in the several jails of the Province'. After reading *Finger Prints*, Henry's office worked on a hybrid of the Galton and Bertillon systems: using measurements to locate an individual record his identification technique relied on matched thumb-prints to establish individual identity. But establishing a consistent set of Bertillon measurements across 120 different sites in Bengal remained a serious, and intractable, problem.[12]

After initiating correspondence with Galton and visiting his London laboratory in 1894 Henry was convinced that fingerprinting could eliminate the errors produced by his operators, and significantly reduce the cost of identification by dispensing with Bertillon's expensive brass instruments. Two of his subordinates, Azizul Haque and Chandra Bose, began to research a workable system for the classification of fingerprints. The Henry System that emerged from this collaboration made significant changes to Galton's classification, replacing his three types (Arches, Loops and Whorls) with just two (Loops and Whorls) and classifying the fingers in five paired sets, each of which could be one of four possible classifications. Henry's system allowed for a basic classification of 1,024 possibilities and these could be neatly arranged into a cabinet that had 32 rows and 32 columns. For further sub-classification his system relied on the uncontroversial procedures of ridge-tracing and ridge-counting. The labels generated under his system could be neatly represented mathematically, and they seemed to do away with the difficult interpretative problems of Galton's sub-classifications. From the middle of 1897 this system of identification became mandatory for the processing of prisoners across the vast territory of British India. The huge subject population covered by the new system soon conferred on Henry's system the mantle of practical success.

In 1900 Henry published the workings of his system in *Classification and Uses of Finger Print*, the definitive manual of twentieth-century fingerprinting, which described practical methods for what we now call one-to-one and one-to-many matching.[13] He was called to London to advise the Belper Committee on the introduction of fingerprinting in

---

[11] Sengoopta, *Imprint of the Raj*, 128.    [12] *Ibid.*, 130–4.
[13] Cole, *Suspect Identities*, 81–7, Sengoopta, *Imprint of the Raj*, 138, 205–16.

Britain. It was this little book, with its clear explanation of the two different methods of fingerprint identification – that became the agent of a new global system of identification. Galton, as ever, was the energetic advocate of its application across the empire.

In July 1900 Galton published an essay in *Nineteenth Century* which summarised Henry's description of the administrative uses of fingerprinting in India and Egypt where the 'natives are too illiterate for the common use of signatures'.[14] The essay showed that single-print fingerprinting had been adopted across the bureaucracy in India – in the payment of government pensions, document registration, opium advances, medical examinations, plague regulations and migration. But he also complained about the limits of one-to-one matching, signalling the ambition that would distinguish the state's project in South Africa, observing that it was 'a pity to print from the thumb alone' when even three fingerprints would support the development of registers of identification.[15] Galton concluded his essay by predicting what was about to begin on the Witwatersrand. 'It will be a real gain if these remarks should succeed in impressing the public with the present and future importance of Identification Offices, especially in those parts of the British Empire where for any reason the means of identification are often called for and are not unfrequently absent', he wrote. 'I think that such an institution might soon prove particularly useful at the Cape.'[16]

## Scientific identification

It was Joseph Chamberlain, the Secretary of State for the Colonies and the outstanding advocate of a racially-bounded social imperialism, who sent Henry to the Witwatersrand.[17] The despatch that Henry carried to the South African High Commissioner, Sir Alfred Milner, specifically noted that he would have sweeping authority in the building of the new police force, assuring him a 'free hand in his own Department' and it explained that his primary task would be to 'introduce a scientific system of identification'.[18] In sending the leading British expert on criminal identification to the Rand, Chamberlain had in mind a very specific remedy to the problems facing the gold mining industry.

[14] Francis Galton, 'Identification Offices in India and Egypt', *Nineteenth Century* 48 (July 1900): 118–26.
[15] *Ibid.*, 122.   [16] *Ibid.*, 126.
[17] Bernard Semmel, *Imperialism and Social Reform: English Social-Imperial Thought, 1895–1914* (London: George Allen & Unwin, 1960), 26.
[18] Chamberlain, Secretary of State, London to Milner, High Commissioner, Cape Town, 18 July 1900, TAD PSY 1 HC102/00 High Commissioner: Johannesburg Police Henry Appointed Temporary 1900-1900.

Edward Henry was not in South Africa for very long – he remained on extended leave from the Indian Colonial Service for just ten months. He arrived in July 1900 and returned to Britain to take up the post of Assistant Commissioner, Head of the CID at Scotland Yard, and heir apparent to the London Metropolitan Police, in May 1901.[19] In a little more than six months he implemented a set of policies that were derived from the demands that the leading engineers of the gold mining industry had presented to a commission of enquiry hosted by the Boer Republic a few years earlier. Bound and typeset by the Chamber of Mines in the fashion of a Parliamentary Blue Book, the 1897 Mining Industry Commission provided a meticulously documented long-term plan for the implementation of the new mining corporations' requirements of the Transvaal state.[20]

Chief amongst these was the mining corporations' demand for a system of identification to bind migrant labourers to the terms of their long-term contracts. 'If every kaffir could be traced; if it could be told whether they have been registered before, or been in the service of a company', one of the leading American engineers had explained to the Transvaal commissioners, 'then we would have control over them'. The Boer officials, unaware or sceptical of the claims of Bertillon and Galton, had answered this demand by explaining that 'when a native throws away his passes and badge he cannot again easily be identified by pass officials', and they had challenged 'anyone to describe a native, and register him in such a way that he would be able to identify him without the aid of his passes and badge, out of 60,000 other natives'.[21] It was this apparently magical project that drew Edward Henry to the Witwatersrand.

But the making of a system of identification was not his only objective. Shortly after he arrived on the Witwatersrand Henry was given the much bigger job of building the entire police force from scratch, and here again he took his instructions from the report of the 1897 Commission. During his time in South Africa Henry maintained close personal links with both Milner and Lord Roberts, the overall commander of the British forces. He left Cape Town for Johannesburg on 12 August with Milner's warm endorsement.[22] In less than a month on the Rand he produced a plan for

---

[19] G. Dilnot, *The Story of Scotland Yard* (Boston: Houghton Mifflin Company, 1927), 134; D.G. Browne, *The Rise of Scotland Yard* (London: George G. Harrap, 1956), 269.

[20] Witwatersrand Chamber of Mines, *The Mining Industry, Evidence and Report of the Industrial Commission of Enquiry, with an Appendix* (Johannesburg: Chamber of Mines, 1897).

[21] *Ibid.*, 44, 279.

[22] Milner, High Commissioner, Cape Town to Lord Roberts, Pretoria, 'Telegram', 12 August 1900, TAD PSY 1 HC102/00 High Commissioner: Johannesburg Police Henry Appointed Temporary 1900–1900.

the establishment of the Johannesburg police that had the enthusiastic support of both the Proconsul and the Field-Marshall.

Henry was different from the small group of young, Oxford-educated disciples that Milner gathered around him.[23] He was an experienced colonial administrator, and already something of a metropolitan celebrity – not least because he had eclipsed Galton in the public hearings of the Belper Commission.[24] Within weeks of his arrival on the Rand he had the fervent backing of Lord Roberts for his plan and – at a time when the government and the army were at continuous loggerheads – he kept Milner on side with a string of hand-written private letters. Both men agreed that 'Henry would be [the] best' appointment as the new Transvaal Commissioner of Police if he could only be persuaded to accept a salary of £2,000.[25] Henry turned down the offer, but he was clearly tempted, explaining coyly to Milner that 'I quite realise that the pay offered is as large as the financial problem justifies, and I have no doubt that a suitable man will be obtainable somewhere'.[26] And Milner, especially, relied very heavily on Henry in the design of the new state. Towards the end of 1900 he wrote to Chamberlain to ask for his support in securing an extension of Henry's leave from the Indian Colonial Service. 'He is doing most important work which no one else can do as well', Milner explained, 'and it seems to me absolutely necessary in the public interest that we should keep him in the Transvaal for some time longer'.[27]

This job, which Henry mostly made for himself, was to mark out the foundations of the unusual national police force in South Africa. Between July 1900 and March 1901 he set in place a police structure that broke completely with the international norm of the city as the administrative and geographical platform of policing. Beginning with the problem that Johannesburg – then as now – attracted 'the most notorious and skilful criminals in Europe', Henry proposed a force of some 400 officers

---

[23] Walter Nimocks, *Milner's Young Men: The 'Kindergarten' in Edwardian Imperial Affairs* (London: Hodder & Stoughton, 1970); Donald Denoon, *A Grand Illusion: The Failure of Imperial Policy in the Transvaal Colony during the Period of Reconstruction, 1900–1905* (London: Longman, 1973).

[24] Sengoopta, *Imprint of the Raj*, 180–1.

[25] Milner, High Commissioner, Cape Town to Lord Roberts Pretoria, 'Telegram', 8 October 1900, TAD LD 15 LAM 1150/01 Commissioner of Police Reports Transvaal Police 1901.

[26] Edward R. Henry to Milner, High Commissioner, Cape Town, 11 October 1900, TAD LD 15 LAM 1150/01 Commissioner of Police Reports Transvaal Police 1901.

[27] Milner, High Commissioner, Cape Town to Chamberlain, Secretary of State, London, 'Telegram', 20 December 1900, TAD LD 15 LAM 1150/01 Commissioner of Police Reports Transvaal Police 1901.

reporting to a single commissioner, a plan modelled on the London police. As his responsibilities grew to include Pretoria, and then out into the countryside, his arrangements for a single unified police force followed. From an initial proposal for a peacetime Johannesburg police force his plans expanded in the Transvaal Town Police Act of 1901 to include all of the towns of the old Republic, with Inspectors in Johannesburg and Pretoria and Superintendents in four of the large towns, all reporting to a single Commissioner of Police in the capital.

This unusual structure of policing – which today requires individual stations to report to the national headquarters, usually many hundreds of kilometres distant – provided a very early and muscular push for the concentration of power in Pretoria. Merriman, the last prime minister of the old Cape Colony, aptly described it in 1915 as a 'highly centralised little army'.[28] But there is no evidence in Henry's correspondence that this centralisation was intentional; it was, rather, a result of the combination of the proximity of the only two large towns, the South African lawyers' interest in preserving the Boer political tradition and the British-led state design that Milner fostered after 1900.[29] Local municipal governments had, under the conditions of imperial reconstruction, no opportunity to build their own police forces, and after the restoration of white self-government the leaders in Pretoria consistently mistrusted the elected leaders of the municipalities.

### Drink, gold and passes

Henry's design for the Transvaal Police was organised around three unusual laws that had been drawn up in the old Republic in the aftermath of the 1897 Mining Industry Commission. Most of this work was

---

[28] Edward Henry to Sir Alfred Milner, 'Memorandum on a Police Organisation for Johannesburg and the Witwatersrand District', 18 September 1900, TAD LD 15 LAM 1150/01 Commissioner of Police Reports Transvaal Police 1901; Edward Henry to Fiddes, W, 'Police Organisation for the Towns of the Transvaal', 2 November 1900, TAD LD 15 LAM 1150/01 Commissioner of Police Reports Transvaal Police 1901; Edward Henry to General Maxwell, 'Proposal for the Policing of Pretoria', 26 November 1900, TAD LD 15 LAM 1150/01 Commissioner of Police Reports Transvaal Police 1901; Edward Henry to Sir Alfred Milner, 'Draft Act for the Regulation of the Town Police', 11 November 1900, TAD LD 15 LAM 1150/01 Commissioner of Police Reports Transvaal Police 1901; see Annette Seegers, 'One State, Three Faces: Policing in South Africa (1910–1990)', *Social Dynamics* 17, no. 1 (1991): 36–48; and Martin Chanock, *The Making of South African Legal Culture, 1902–1936: Fear, Favour and Prejudice* (Cambridge University Press, 2001), 46–9 for discussions of the Transvaal Police as the foundation of the Union police force. The quote is from J.X. Merriman, cited on 49.

[29] Henry to Milner, 'Memorandum on a Police Organisation'; Edward R. Henry to Fiddes, W, 13 February 1901, TAD LD 15 LAM 1150/01 Commissioner of Police Reports Transvaal Police 1901.

undertaken by the young Cambridge-educated Secretary of State, Jan Smuts, under pressure from the American engineers who dominated the gold mining industry after 1893. The first was Smuts' Liquor Law (Act Number 19 of 1898) which – after more than a decade in which African mine labourers had been encouraged by their bosses to drink away their wages – imposed a sweeping and fierce prohibition on the sale and consumption of alcohol on anyone with a dark skin.[30] The Liquor Law, one of only three that Henry included in the standing instructions of the new police force, prohibited the sale of alcohol on mining ground and to any 'coloured person', defined in sweeping terms to include any 'African, coloured American, Asiatic, Chinaman or St Helena person'.

The second preoccupation of Henry's new police force was something called the Gold Law, a massive document that provided very generous legal support to the mining companies in dozens of areas (the law took up nearly 100 printed pages). Again following the demands to the 1897 Commission, Henry's police force took the enforcement of the provisions that dealt with the sale and possession of unprocessed gold as a foundational task – prohibiting trade in 'unwrought precious metal' and establishing the presumption of theft and harsh penalties (especially for Africans) for 'anyone found in possession of amalgam or unwrought gold'.[31]

These two little pockets of law geared to the requirements of the gold mines – the prohibition of alcohol for Africans and the criminalisation of the simple possession of unwrought gold – each provided the focus for a special-purpose branch of Henry's new Criminal Investigation Department.[32] For years after his departure fully two-thirds of the investigative capacity of the new police force was dedicated to the peculiar security requirements of the mining industry. Henry's CID was made up of an underfunded Headquarters section, which had responsibility for his identification registers and crime that did not involve the mines, a Gold Branch – with special funds for the employment of private detectives to

[30] On Smuts' efforts to reform the Transvaal see Charles van Onselen, *Studies in the Social and Economic History of the Witwatersrand, 1886–1914: Volume 1 New Babylon* (Johannesburg: Ravan Press, 1982), 14, and 44–96 for the longer story of prohibition. Also C. van Onselen, *The Fox and the Flies: The World of Joseph Silver, Racketeer and Psychopath* (London: Jonathan Cape, 2007), 161–80 on Smuts' efforts to reform the anaemic Johannesburg detectives before 1900. For the ongoing importance of alcohol for mine workers, Patrick Harries, *Work, Culture and Identity: Migrant Labourers in Mozambique and South Africa, c 1860–1910* (Johannesburg: Witwatersrand University Press, 1994), 194–200.

[31] Chamber of Mines, *The Mining Industry, Evidence and Report of the Industrial Commission of Enquiry, with an Appendix*, 641.

[32] Henry to Milner, 'Memorandum on a Police Organisation'.

trap dealers and mine employees tempted by the illicit trade in precious metals – and a Liquor Branch. Before Apartheid, it was the prohibitionist arm of the police force that cut most deeply into the daily lives of black urban residents, with hundreds of thousands of men and women convicted and imprisoned every year.[33]

Another result of this peculiar division of the state's policing resources was very weak capacity for the investigation of what Henry called Ordinary Crime. At the outset, to save money, Henry did not fill the detective posts in his Headquarters section; a few years later, in a move that reflected the melancholy standing and capacity of the Ordinary Crime branch of the CID, it was, temporarily, dissolved altogether.[34]

The result was a paradox: the police force designed by the most famous English-speaking advocate of forensic policing[35] was bereft of investigative capacity in general and a working detective CID in particular. Henry designed the police force around the particular needs of the mining industry, with little regard, at all, for the society that was growing up around it. The allocation of the lion's share of the detective capacity to the mines was one aspect of this special policing apparatus. The enforcement of the Pass Law was another. In his first draft of the blueprint for the Transvaal Town Police, Henry carefully set up the distribution of personnel. Aside from the 'more serious offenses against the Penal Code', he explained, the police force 'will have to occupy itself specially with the detection of Gold thefts and violations of the Liquor Law' and it 'would be the agency mainly utilised with the working of the pass law'.[36]

It was this focus on the Pass Law, and particularly the provisions of a new law drafted in the last weeks of the Boer Republic, that set in place the engine of incarceration in South Africa. Henry's standing instructions, which were published as a pamphlet in 1901, included the verbatim text of both the General Pass Law of 1895 and Smuts' carefully refined Act No. 23 of 1899. This last law, drafted by Smuts but never

---

[33] For an overview of this literature, see J.S. Crush and Charles H. Ambler, *Liquor and Labor in Southern Africa* (Athens, OH: Ohio University Press, 1992).

[34] John A. Bucknill to Attorney General, 'Report of a Departmental Enquiry Held by Mr John A Bucknill into Certain Complaints Made by the Commissioner of Police against the Acting Chief Detective Inspector on His Establishment and into Other Matters in Connection Therewith', 7 May 1906, TAD C30 11 PSC 454/06 Mr Brucknill's report on certain police matters 1906; for more on the anaemic detective resources in the Transvaal after Henry, see Charles van Onselen, 'Who Killed Meyer Hasenfus? Organized Crime, Policing and Informing on the Witwatersrand, 1902–8', *History Workshop Journal* 67, no. 1 (20 March 2009): 1–22.

[35] Sengoopta, *Imprint of the Raj*, 183. Admittedly, even after Henry, Scotland Yard was notably less enthusiastic about scientific policing than its European peers or the FBI, see Browne, *The Rise of Scotland Yard*, 206.

[36] Henry to Milner, 'Memorandum on a Police Organisation'.

properly implemented because of the impending war, adopted the mine engineers' demand that black men found on the Witwatersrand without passes should be subject to a week-long period of imprisonment to allow for the identification of deserters.[37] In the decades that followed it was this principle – the normal practice that African men (and eventually women) found on the Witwatersrand without an appropriate pass were subject to at least a week of compulsory detention to enable the state to identify them – which became the distinguishing characteristic of the South African state.

There are two pieces of history here that have not been noticed before. The first is that automatic imprisonment for violation of the pass law was Smuts' innovation, and that it dates from the last days of the Boer Republic (not the Native Labour Regulation Act of 1911). This is not surprising but it adds to the large body of evidence of his protean significance in the fashioning of the twentieth-century state. The second is Henry's careful, and deliberate, adoption of Smuts' law to initiate the modern pass regime in South Africa.[38] This speaks to the problem of how the empire should be digested by historians of the comparatively restrained forms of state power (relative to Germany in particular) that developed in Britain in the twentieth century.[39] The

---

[37] 'TAB SS 0 R13775/98 Publieke Aanklager Pretoria. Geeft Wyziging Aan Den Hand Van Wet 2/1898 (paswet). 18981019 to 18981019 See R5353/98.', n.d., www.national.arch-srch.gov.za/sm300cv/smws/sm30ddf0?200806291653545A3C588C&DN=00000082; 'TAB SP 177 Spr6780/98 Kamer Van Mynwezen Johannesburg. Re Detentie Van Naturellen Voor Identificatie. 18980715 to 18980715', n.d.; Edward Henry, *Police Regulations, & c: Published by Authority* (Pretoria: Government Printing Office, 1901).

[38] Harries, *Work, Culture and Identity*; Douglas Hindson, *Pass Controls and the Urban African Proletariat* (Johannesburg: Ravan Press, 1987); Alan H. Jeeves, 'The Control of Migratory Labour on the South African Gold Mines in the Era of Kruger and Milner', *Journal of Southern African Studies* 2, no. 1 (1975): 09/03/29; E. Kahn, 'The Pass Laws', in *Handbook of Race Relations in South Africa* (Johannesburg: Race Relations Institute, 1949), 275–91; Michael Savage, 'The Imposition of Pass Laws on the African Population in South Africa 1916–1984', *African Afairs* 85, no. 339 (1986): 181–205; Charles van Onselen, 'Crime and Total Institutions in the Making of Modern South Africa: The Life of "Nongoloza" Mathebula, 1867–1948', *History Workshop Journal* 19 (1985): 62–81; Peter Warwick, *Black People and the South African War (1899–1902)* (Cambridge University Press, 1980). Almost all of these scholars have relied on Van der Horst's study, which jumps from the revised 1895 law to Lagden's Proclamation Number 37 of 1901, see Sheila T. van der Horst, *Native Labour in South Africa* (Oxford University Press, 1942); for the most comprehensive and recent overview of the Pass Law see Chanock, *The Making of South African Legal Culture*, 406–22. Chanock also notes Edward Henry's appointment on the Rand, see 46.

[39] Edward Higgs, *The Information State in England: The Central Collection of Information on Citizens since 1500* (New York: Palgrave Macmillan, 2004); Jon Agar, *The Government Machine: A Revolutionary History of the Computer* (Cambridge, MA: MIT Press, 2003); Jon Agar, 'Modern Horrors: British Identity and Identity Cards', in *Documenting Individual Identity: The Development of State Practices in the Modern World*, ed. Jane Caplan and John

global political effects of what Hannah Arendt called colonial despotism were more paradoxical than her account of the movement from the British empire to Nazi Germany allows.[40] Henry was one agent, a very important one, in the long evolution of the systems of incarceration that would imprison nearly 10 per cent of the adult population every year in South Africa in the last century. But perhaps the most meaningful result of this work was that the forms of identification he championed were tainted by their association with the coercive conditions on the Witwatersrand.

A system for the imprisonment of undocumented black workers would have developed in the Transvaal had Edward Henry not resuscitated it, if only because Milner's legal advisers – Richard Solomon and John Wessels – were determined to build the legal order of Milner's new state on the foundations of the old Republic.[41] Smuts' law already existed. The same is not true of the way in which Henry built fingerprint identification into the core administrative practices of the new state beyond the Pass Law. Henry's special contribution to the twentieth-century state was to weave an obsessive preoccupation with fingerprinting as the only reliable means of identification into many different levels of the new bureaucracy. It was also Henry who introduced the long-running fantasy that only a single, centralised register of the identities of Africans could solve the problems of impersonation and uttering that were prompted by the coercive demands of the Pass Laws.

When Henry described the work of the Criminal Investigation Department in 1900, he had himself in mind as its first commissioner. And he set himself the first practical task to 'introduce and keep under his management a scientific system of identifying old offenders as an aid to criminal investigation'.[42] After he left the Witwatersrand the Records Section was the one undisputed success of the CID.[43] He had left it in the care of a young and ambitious Cypriot detective recruited from the Egyptian police. In the decades that followed the CID fingerprint

Torpey (Princeton University Press, 2001), 101–20; see Edward Higgs, 'Fingerprints and Citizenship: The British State and the Identification of Pensioners in the Interwar Period', History Workshop Journal 69 (2010), for a discussion of the political implications of Empire for systems of surveillance in Britain.

[40] Hannah Arendt, The Origins of Totalitarianism (San Diego: Harcourt Brace and Company, 1973).

[41] Richard Solomon to Fiddes, W, 2 February 1901, TAD LD 15 LAM 1150/01 Commissioner of Police Reports Transvaal Police 1901; Solomon, Richard, 'Enclose in No 15. Minute. Penalties under the Pass Law and Gold Law of the Transvaal', 15 July 1901, Cd 904 Papers relating to legislation affecting Natives in the Transvaal (in continuation of [Cd 714], July 1901).

[42] Henry to Milner, 'Memorandum on a Police Organisation'.

[43] Bucknill to Attorney General, 'Bucknill Enquiry into Certain Complaints', 11, 28.

department assumed a position of contested authority over the many other fingerprint registries. It was also the most influential source of the argument for universal registration and the institution that other colonial governments in Africa consulted for plans and training in the development of their own systems of fingerprint identification. But the police fingerprint branch was less significant in the long run than the wider philosophy of policing that Henry articulated in 1900.

Working in concert with the CID fingerprint department Henry produced and published a pamphlet on the standard operating instructions for the Transvaal Police in 1901. This booklet, titled *Police Regulations* and modelled on the General Instruction Book that Sir Richard Mayne had drawn up for the London Metropolitan Police half a century before, established a policing common sense that remained consistent until the end of the Apartheid state.[44] Like Mayne's book, Henry provided a list of the laws which he thought every policeman in the Transvaal should know. These were, as we have seen, the 1895 and 1899 Pass Laws and the Liquor Law of 1898. Where Mayne's book (which remains in use in Britain) established a philosophy of police restraint by elaborating on the 'preventive idea in a manner allowing little misunderstanding',[45] Henry's pamphlet enshrined the idea that Africans, because they could not be trusted to identify themselves, should routinely be subjected to police detention.

In addition to highlighting the Pass and Liquor Laws, Henry's new Code specifically instructed his officers to apply a very different test of identity, and detention, to Africans, because of their 'detachable names'. This instruction appears twice, and in some detail, in the Code. First, in the explanation of the appropriate conduct for police on the beat:

Generally it must be understood that as Natives have not a permanent residence within the limits of the Police District, and are known by detachable names, such as 'Charlie,' 'Jim,' etc other than their proper tribal name, and in consequence cannot be easily traced, it is often necessary to arrest them in cases in which, if the accused were European, another course might be preferred.

And then again in the advice on the procedures for the granting of bail, he combined English restraint with imperial discipline:

The Police should understand that all unnecessary confinement in cells is to be avoided, but in the case of Natives, the majority of whom have not a permanent residence within the limits of the Police District, and who are known by detachable names which makes it difficult to again trace them, bail should

---

[44] Henry, *Police Regulations, & c: Published by Authority.*
[45] Browne, *The Rise of Scotland Yard*, 78; Dilnot, *The Story of Scotland Yard*, 45.

be more sparingly accepted than in the case of Europeans, and their personal recognisance should very rarely be accepted.[46]

Combined with the Liquor and Pass Laws, it was this presumption of universal deceit that set the technology of fingerprint identification into the foundations of the South African state.

Henry's most important efforts were not confined to the finger-print department of the police. From very early on he worked to build fingerprint registration into the operations of the Pass Offices on the Witwatersrand. In October, his second month on the Rand, he wrote to Milner to explain that he had started 'the F Printing system of identifi-cation & some of the men here are becoming proficients'. The problem on the Rand, he explained, lay in the fact that black workers were being given names by their employers, who evidently had little interest in, or capacity for, understanding their workers' real names. 'Every Native appears to be called Charlie or Sammy or Jonnie or some such English name', with the unfortunate result that 'when he changes master he often changes his name'. Implicitly comparing his fingerprinting system to the mandatory birth, death and marriage registration that was introduced by Henry VIII in 1538 he complained that 'just as it was found impossible in England five or six hundred years ago to describe persons correctly in writs or other legal documents though their persons were known so it is here'.[47]

Henry's solution to this problem of 'detachable names' was to build an official identification register out of the special labour regulations demanded by the mining industry. 'To effect reform it is necessary to start with the Pass Offices and this is now being done', he told Milner. He had in mind the careful recording of the real names of every African man on the Rand, 'not his phony appellation while working here'. And to counter the problems of recording 'Zulu, Basuto etc names' he explained that he would build 'a system of indexing which to some extent discounts the effect of variations in transliterating'.[48] From the Pass Offices he

---

[46] Henry, *Police Regulations, & c: Published by Authority*. For Henry's drafts of this document see TAD LD 15 LAM 1150/01 Commissioner of Police Reports Transvaal Police 1901.

[47] Henry to Milner, High Commissioner, Cape Town, 11 October 1900. On the signifi-cance of the Cromwellian registration for the identification of the English, see Simon Szreter, 'Registration of Identities in Early Modern English Parishes and Amongst the English Overseas', in *Registration and Recognition: Documenting the Person in World History*, ed. Keith Breckenridge and Simon Szreter, Proceedings of the British Academy 182 (Oxford University Press, 2012), 67–92; Edward Higgs, *Identifying the English: A History of Personal Identification, 1500 to the Present* (London and New York: Continuum, 2011), 79–81.

[48] Henry did not elaborate on the workings of this system, but he may have been referring to the method of fingerprint classification and indexing he described in Part II of his

hoped that the properly recorded names would flow into the police sys-tem, providing accurate identification data in the new Criminal Registers of the CID. For the documents themselves Henry suggested that a single 'thumb or digit mark on the Pass would ensure that these passes are used by those only to whom they were issued'.[49]

In fact the situation that Henry left on the Witwatersrand was richly contradictory. When the Native Affairs Finger Impression Record Department started working in October 1901 – six months after Henry had returned to England – they duly began to collect fingerprints from the tens of thousands of men who had applied for permission to work on the Rand. But the system of indexing Henry had described soon proved much too difficult, and much too expensive, to implement. The result was that only the fingerprints of workers who had been convicted of a crime, or those whose employers had reported them as deserters, were classified according to Henry's system, and filed in a special register of suspects.[50] A much larger group, soon running into hundreds of thousands, were registered by name and passport number; their unclassified fingerprints were accessible only by their passport numbers. But all workers who were caught by the police without a properly endorsed pass 'or those whose passes or papers are in any way doubtful' were, nonetheless, auto-matically subjected to a week of imprisonment to give the Pass Officials time to look up their fingerprints against the register of suspects – a pro-cedure that rarely worked as it was designed.[51] This banal and pervasive imperative for imprisonment – which hovered around one-tenth of the adult male black population every year up to the 1960s and then grew to include a similar proportion of women – was another distinctive feature of the South African state in the twentieth century.[52]

book. See E.R. Henry, *Classification and Uses of Finger Prints* (London: G. Routledge and Sons, 1900), 66–90.

[49] Henry to Milner, High Commissioner, Cape Town, 11 October 1900.

[50] Wilson, Edward, Superintendent FIRD, Pass Office, Johannesburg to District Controller, 2 May 1906, TAD SNA 332 Na2491/06 Part 1 Proposed Extension of Finger Impression Record Department Na2782/06 Part 1 the Under Secretary Division 2 Colonial Secretary's Office Finger Print Impression Experts 1906; Edward Wilson, Superintendent FIRD to H S Cooke, Acting Pass Commissioner, 'Report Working of the NAD Pass Office FIRD', 12 February 1907, TAD SNA 354 The Acting Pass Commissioner, Johannesburg, forwards copy of a report by the superintendent, Finger Impression Record Department, 1907.

[51] Wilson, Edward, Superintendent FIRD, Pass Office, Johannesburg to District Controller, 2 May 1906.

[52] For discussion of the statistics see Savage, 'The Imposition of Pass Laws on the African Population in South Africa 1916–1984'; for estimates of adult males see time series in Pali Lehohla, *South African Statistics 2009*, Annual Report, South African Statistics (Pretoria: Statistics South Africa, 2009).

Henry's influence in South Africa was not limited to the design of the Transvaal Police or the Pass Law. The most enthusiastic disciples of fingerprinting, after 1900, were in the two adjacent colonies that would become provinces in 1910. In Natal, the man who would later become Commissioner of the Provincial Police, William James Clarke, had attempted to persuade the impoverished colonial legislature to set up an identification bureau in the late 1890s. Like many of the key sites of biometric registration, the office in Natal was the life's work of a single determined individual. When Clarke's proposal was denied, he used his own money to set up an office of Bertillonage. In 1902 he used his annual leave (and one of his most ambitious NCOs) to visit Henry at the finger-print offices of the Metropolitan Police. From this point the Natal branch of the CID became a site of frenzied fingerprint registration. Clarke's branch collected prints from all indentured Indians arriving in Durban, and all Africans who interacted with the colonial state: those convicted of criminal offences (including the Natal versions of the Pass Law) and those who sought employment in the Police and Prisons Service. By 1907 the Natal CID had a collection of 160,000 prints, which, as Clarke liked to point out, meant that his 'system is more complete than Scotland Yard'. Over the next two decades he would remain a tenacious advocate of the virtues of biometric registration (see Chapter 4). When the little colony joined the Union in 1910 it brought in to the new police force a collection of 350,000 prints organised according to the Henry System, out of a total population of slightly more than one million people.[53]

In the years after 1902 William Clarke claimed to be in 'constant communication' with Scotland Yard, but that was hardly necessary. The spirit of Edward Henry remained a powerful presence long after he had left South Africa. Much of this influence was enacted through circulation of the different editions of his book, which were ordered by the boat-load by many different divisions of the state. The book was also dissolved into the correspondence of the bureaucracy itself. When the Secretary for Law in the Transvaal explained the importance of fingerprints to the new magistrates in 1905 his memorandum consisted of six lengthy extracts from Henry's book. After noting that Galton had proved the life-long persistence of fingerprint patterns, the circular repeated Henry's claims that fingerprints were an 'extraordinarily efficient method of preventing perjury

---

[53] See Clarke's comments on N.P. Pinto-Leite, 'The Finger-Prints', *The Nonqai* (1907): 31–3; and 'NAD MJPW 104, LW5435/1903 Agent General, London, Forwards a Letter from the Right Honourable Sir H. Macdonell with Reference to the Prospects of Mr. Pinto Leite Obtaining a Commission in the Natal Police', 1903; Truter, Theo G. Acting Chief Commissioner of the South African Police to Acting Secretary for Justice, 'Finger Print System in South Africa', 27 May 1912, CAD JUS 0862, 1/138, URL.

and personation' because the distinctive ridge patterns 'differentiate each individual from all others'. As an 'instance of the practical use of the system' the memorandum included a verbatim account of the 'notorious criminal case' that concludes the first part of Henry's book – the murder investigation of the Julpaiguri tea garden manager. The concluding paragraph quoted from the highlights of the 1897 Indian Government Committee Report that Henry had appended to his book. 'The greatest sceptic would be at once convinced of identity on being shown the original and duplicate impressions. There is no possible margin of error and there are no doubtful cases.'[54]

The main focus of the fingerprinting enterprise in South Africa before 1907 was on the effort to control indentured Asian workers, and particularly the very large numbers of Chinese men contracted to work in the Witwatersrand gold mines. Chinese workers presented the officials of the Foreign Labour Department (FLD), and their counterparts on the individual gold mines, with formidable problems of identification and regulation: there was practically no proficiency on the Rand in any of the languages spoken by the migrants; many of the workers had similar or identical names; the officials were completely incapable of distinguishing individual physical characteristics; and, simply put, workers had many good reasons to dissemble about their own identities.

The FLD initially sought to address the problem of establishing reliable personal identity by using a new system of numerically indexed photographs. The first group of workers arrived at the Receiving Compound built on the site of the old Boer Concentration Camp at the Jacobs Railway Depot in Durban, after a thirty-day journey from Hong Kong. Officials took photographs of each worker holding a slate with the number of the Government Passport issued to him immediately before the departure of the ship. The photographs travelled with the workers on the twenty-four-hour train journey to the Transvaal where they were carefully filed in numerical order in the FLD's Identification Office in Johannesburg.[55] But these images proved singularly incapable of the work required of them. The rapid processing demanded for the 2,000 men aboard each ship meant that many of the photographs were in fact useless for the purposes of identifying individuals. And even when, by chance, the pictures

---

[54] Acting Secretary to the Law Department, 'Finger Prints as Evidence of Identification', 13 March 1905, TAD LD 1035 1193-05 Finger prints as evidence of identification 1905; for a critical study of these claims see Cole, *Suspect Identities*, 190–215; and Simon A. Cole, 'Witnessing Identification: Latent Fingerprinting Evidence and Expert Knowledge', *Social Studies of Science* 28, no. 5/6 (1998): 687–712.

[55] Peter Richardson, *Chinese Mine Labour in the Transvaal* (London: Macmillan, 1982), 146–9, 160–5.

were well taken, photography was of little help in presenting the officials with a set of easily distinguishable physical criteria. And, 'when they ran into thousands', as the officials put it, they had little hope 'to survive the ordeal of constant handling'.[56]

Following the example of the Paris police, to bolster the flimsy evidentiary qualities of the photographs the FLD officials in China first turned to Alphons Bertillon's 'speaking portrait' which 'combined photographic portraiture, anthropometric description, and highly standardized and abbreviated written notes on a single *fiche* … within a comprehensive, statistically based filing system'.[57] This system, which must have seemed omniscient and elegant on display at the 1893 Chicago Exposition, simply added to the problems of identifying large numbers of workers on the Witwatersrand. Far from resolving the empirical inadequacies of the photographs, Bertillon's complex measuring instruments demanded more time and more skill than the FLD could muster. Nor did this effort 'to ground photographic evidence in more abstract *statistical* methods' do much to address the intrinsic unreliability of the entire body as a point of mathematical comparison. The measurements, as officials complained at the time, were simply 'not reliable owing to physical changes'.[58]

Fingerprints, as Galton had argued some twenty years earlier, proved significantly more efficient – and much cheaper – than Bertillon's anthropometrics.[59] Following the methods that Henry had left behind, the FLD officials in Johannesburg began to collect, classify and file the fingerprints of the Chinese migrants as they arrived in the city. Within a year they had collected some 13,000 sets, and were steadily increasing the catalogue at a rate of 4,000 fingers per day. Building, classifying and filing this mountain of identifying data required the labour of just two registrars at the receiving compound and three cataloguers at the FLD's Identification Office in downtown Johannesburg.[60]

Fingerprinting the Chinese workers offered the state one particular advantage: a racially delimited record set that was limited and relatively easy to complete. Within five years every single Chinese worker – indeed eventually every Chinese or Indian adult male in the Transvaal – was

[56] Burley, Henry, to Registrar of Asiatics, Department of the Interior, 'Report on the Workings of the Fingerprint System', 22 April 1912, CAD JUS 0862, 1/138, URL.
[57] Sekula, 'The Body and the Archive', 18.
[58] Burley, Henry, to Registrar of Asiatics, Department of the Interior, 'Report on the Workings of the Fingerprint System'.
[59] Galton, *Finger Prints*, 166–8.
[60] Identification Branch, Foreign Labour Department, 'Report on the Work of the Identification Branch of the Foreign Labour Department', 1 September 1905, TAD FLD 171, 35.

fingerprinted. Similarly, while the total figure of 70,000 men was large, it was small compared to the millions of potential African workers.

The FLD was able to maintain the size of the record set by organising all aspects of the recruitment and control of the Chinese workers – including the fees and wages charged to the mines – around their fingerprints. From early in 1905 the mines were required to dispatch a set of fingerprints along with the notification of death of Chinese workers. The integrity of the collection was nicely maintained by linking the fingerprint records to each company's wage bill. 'Your company will be saved the sum of Ten Shillings per month on the allotment payments', an FLD memo carefully reminded the mines' secretaries, 'which cannot be stopped until the right coolie is reported as dead'.[61]

The fingerprinting of the Chinese workers very quickly came to serve as a model of registration. When, two years after the last of the Chinese workers had been repatriated and the collection had been destroyed, the officials of the FLD (who had now been transferred into a new 'Asiatics' unit within the Interior Ministry) looked back on the Chinese era as a period of faultless state control.

It will of course be understood that the Chinese resorted to all sorts of devices in order to make some attempt to lose their identity ... Passports were exchanged, false names given, the names of employers were wrongly given and in a number of cases even the mutilation of fingers that existed at the time they were originally taken ... It is therefore a tribute to the excellence of the system that during the seven years these 70,000 labourers were employed in the Transvaal there is NO case on record of mistaken identity having occurred!

The Chinese episode seemed to offer disciplinary possibilities that would dissolve the most persistent barriers to the proper organisation of an enormous colonial labour force. These obstacles, first cultural – 'the Chinese aptitude for prevarication' – and, second, linguistic – 'no one of the finger print experts spoke the Chinese language' – collapsed under a regime where officials confronted the migrant exclusively 'through his finger prints and the classification under which they fell'.[62] This triumphalist vision of the Chinese era, drawing on the grandiose tones of Francis Galton's original study, came to underpin the notion – fondly held by compound managers, police commanders and mine managers – that fingerprinting could solve the weaknesses of the documentary Pass system.

There were plenty of problems with the identification of the Chinese – the same that would bedevil fingerprinting later in the century – but the

[61] *Ibid.*
[62] Burley, Henry, to Registrar of Asiatics, Department of the Interior, 'Report on the Workings of the Fingerprint System'.

statistics, at least, were unequivocal. In 1906 the FLD registered and classified in excess of 10,000 newly arrived migrants, they identified 12,000 Chinese workers released from prison, 15,000 appearing in court, 5,000 repatriations and a small number of criminal and miscellaneous identifications, all form fingerprints. By 1910, after the last of the workers had returned to mainland China via Singapore, the collection, which had accounted for the repatriation of every single individual, was completely destroyed. But the legacy of a working example of fingerprinting as an all-inclusive solution to the inadequacies of the documentary order would remain for years.

Henry's influence can also be measured by the reports that were drafted a decade later to answer a request from the Belgian government – themselves busy with the labour regulations for the new Union Minière plant in Elisabethville – on the 'workings of the finger print system in South Africa'. The officials in charge of the large fingerprint repositories in the Native Affairs Department and the Asiatic Registry all proudly acknowledged his paternity. 'The finger print system adopted was that of Sir Edward Henry KCMG, etc, now Chief Commissioner of the London Police', one of the clerks in the Asiatic Registry proclaimed, 'whose method of classification and filing is of almost universal adoption in criminal work and in dealing with Native Labour and Asiatic registration in the South African Union'.[63] The Commissioner of the new Union police explained that the 'finger print system in South Africa is identical with that of Scotland Yard, London, [and] was first introduced by Sir Edward Henry in the year 1901 when he came to the Transvaal to establish the Transvaal Town Police'.[64] The system, he observed, was becoming more valuable 'year by year' and, as a measure of this, he provided a breakdown of the different CID collections in each of the provinces. These repositories, which excluded the massive collections being built up in the Pass Offices on the Rand, showed that in Natal and the Free State nearly a quarter of the African population had already been registered biometrically by the police. But the figures for the Cape showed a paltry 30,000 out of a total population of nearly two million people. There is a significant theoretical point in these differences.

The very small size of the collections in the Cape Province reflect the scepticism that local liberal leaders showed towards the advocates of biometric registration, even in the early period (before – as we will see in the next chapter – Gandhi had given them reason to be sceptical about it). The politics of registration in the Cape was very different from

[63] *Ibid.*
[64] Truter, Theo G. Acting Chief Commissioner of the South African Police to Acting Secretary for Justice, 'Finger Print System in South Africa'.

the institutional regimes at work in the other colonies. The Finger Print Bureau in Bloemfontein, the capital of the Orange River Colony, was built at the same time and by the same imperial officials as the repositories Henry set up on the Rand. After 1906 it was put to use in defence of the Pass Law by the mostly Afrikaans-speaking landowners who led the new *Orangia Unie*. In Natal, where the English-speaking settlers fostered a fiercely coercive racial order modelled on the West Indian plantations, William Clarke's fingerprint bureau was an invaluable cog in machines of indenture and indirect rule.

But in the Cape the remains of a local Gladstonianism, sustained by the presence of African voters, remained powerful into the twentieth century. When the police urged the magistrates in the Transkei (the most recently annexed territories of the Cape Colony governed mostly by a system of indirect rule) to adopt fingerprinting in the gaols, fully half of them announced that they wanted nothing to do with the idea. Despite this opposition the Native Affairs Department recommended that the scheme be adopted because it would be cheap and encourage consistency between the native territories and the Colony Proper. Merriman, the last of the Cape prime ministers, agreed to allow fingerprinting, but his response captures the liberals' indifference to new schemes of government very typically. 'If it only costs £18 I have no objection', he conceded, 'but I am not very sanguine about it'.[65] In contrast, one of the key features of the new states that were being built in the Orange River and the Transvaal Colonies (and even in Natal) at this time was an unrestrained optimism about state-led social engineering.

## Asiatic despotism

Much of this imperial progressivism, and many of the officials, came to South Africa in this period, as Henry did, from India. Like the expert-led planning that fashioned twentieth-century Egypt, powerful schemes of South African social engineering were hatched in Metcalf's 'India-centered subimperial system'.[66] Perhaps the most important of these

[65] Dower, Edward, Secretary for Native Affairs, Cape Colony to Prime Minister, Cape Colony, 'Finger Print System – Habitual Criminals: Proposed Extension to the Transkeian Territories', 17 September 1908, SAB NA 168, 1/1908/F372 Introduction of Finger Print System into the Transkeian Territories, 1913.

[66] Timothy Mitchell, *Rule of Experts: Egypt, Techno-Politics, Modernity* (Berkeley: University of California Press, 2002), 21–42, and on Keynes and India, 83; Robert L. Tignor, 'The "Indianization" of the Egyptian Administration under British Rule', *American Historical Review* 68, no. 3 (1963): 636–61; Thomas R. Metcalf, *Imperial Connections: India in the Indian Ocean Arena, 1860–1920*, 1st edn (Berkeley: University of California Press, 2008), 12.

projects, which would take decades to reach its conclusion, was the plan laid out by William Willcocks for a massive programme of dam building and agricultural irrigation.[67]

The introduction of universal birth and death registration, discussed in Chapter 4 was the result of the Natal government's brief appointment of a Plague Adviser from the Indian Medical Service. When Henry left South Africa his replacement as Commissioner of Police was another veteran of the Indian Colonial Service. Each of these things was to have visible effects on the development of the new state, but it was the technology of fingerprinting that most powerfully joined the two countries together. As we will see in the next chapter, it was out of this connection that Mohandas K. Gandhi emerged as the outstanding figure of Indian nationalism and the most influential twentieth-century critic of what he called 'modern civilisation'.

In his autobiography Gandhi bitterly recalled the 'autocrats from Asia' who had arrived in South Africa with Milner after 1900. Writing in the late 1920s when the bureaucratic culture he was remembering was fast disappearing for black people, Gandhi remembered that 'colour prejudice was of course in evidence everywhere in South Africa' but the local officials 'had a certain courtesy of manner and humility about them'. He contrasted the behaviour of the local officials 'responsible to public opinion' with the bearers of the 'habits of the autocracy' from India. The official he had in mind was Major Montfort Chamney, the man who became the Asiatic Registrar. Chamney, working with Lionel Curtis, was responsible for the elaborate plan for the fingerprint registration of all Asians in the Transvaal. He was certainly an extravagant racist, possessed by the subversive capacities of the Chinese and the Indians, and a relentless advocate of universal biometric registration.[68] In the decade after his arrival in South Africa in 1902 he worked tenaciously to expand Henry's fingerprinting system into an instrument of segregation.[69]

---

[67] W. Willcocks, *Mr Willcocks' Report on Irrigation in South Africa*, BPP Cd 1163 Further Correspondence Relating to Affairs in South Africa. (In Continuation of [Cd 903] January 1902) (Daira Sania Co, Egypt, 1901); the massive project of dam-building in South Africa awaits its history, but see Saul Dubow, *A Commonwealth of Knowledge: Science, Sensibility, and White South Africa, 1820–2000* (Oxford and New York: Oxford University Press, 2006), 259; W. Beinart, *The Rise of Conservation in South Africa: Settlers, Livestock, and the Environment 1770–1950* (New York: Oxford University Press, 2008), 234–65, describes the failed efforts to raise finance for irrigation in the early twentieth century.

[68] In his response to the Belgian government's query in 1912, he suggests, conspiratorially, that the Belgians were interested in fingerprinting because they were intending to employ indentured Chinese workers: Chamney, Registrar of Asiatics, Department of the Interior to Acting Secretary for Justice, 'Confidential', 23 April 1912, CAD JUS 0862, 1/138, 1910.

[69] Montfort Chamney, 'Mahatma Gandhi in the Transvaal', 1935, Mss Eur C859 1902–1914, British Library, Asia, Pacific and Africa Collections, www.nationalarchives.gov.uk/a2a/records.aspx?cat=059-msseur_2&cid=-1#-1.

Chamney followed his uncle, Henry, from the tea plantations of Assam, as part of the much larger migration of Anglo-Indian adventurers who attached themselves to the British Army. Many of these men, including the older Chamney, had volunteered for Lumsden's Horse – a unit of some 250 volunteer mounted infantrymen raised from the imperial regiments in Assam. They were, as the official history of the corps boasted, mostly 'Public School boys' who had 'failed in their examinations for Sandhurst, and gone out to fight their way in India as indigo, tea, and coffee planters'. This combination of middle-class schooling and familiarity with the racial order of the plantations apparently made them 'just the right men to fill the appointments' in South Africa.[70] Lumsden's Horse was the largest single source of recruits for Henry's new police force.

Henry was deeply concerned about the moral qualities of these appointments. He had agreed to take any of the recruits in Lumsden's Horse that the commander might 'specially recommend'. But a month of duty with the Assam volunteers was enough to make him doubt the 'worth of [Lumsden's] recommendations' and the suitability of his soldiers as policemen. After insisting that the recruits had been 'certified to be strictly sober' he was horrified to discover 'one Trooper who it appears is a dipsomaniac'. This disquiet was enough to make him unenthusiastic about Henry Chamney, the most eminent of the Assam volunteers. In his explanation to Milner he drew attention to his own special relationship with the CSI. 'I am an Indian myself and anxious to do a good term to the Indian Contingent', he explained, 'and I think as we have given one Commission and 23 subordinate posts to Lumsden's Horse, they have no grounds of complaint'.

Notwithstanding Edward Henry's doubts, the elder Chamney was appointed as a District Commissioner (in Potchefstroom) in the South African Police. His subsequent career shows the thicket of relationships between the Indian and South African police in this period in sharp relief. After serving a few more years in the new Transvaal Police, he returned to Bengal in 1909 to set up the new Police Training College in Assam. In 1920 he retired from that position, and returned to South Africa to farm in the Rustenburg district on land that he had purchased from Paul Kruger's estate. His nephew remained in the Transvaal throughout this period.

It was Montfort Chamney, son of an Anglo-Irish Doctor of Divinity and erstwhile tea-planter from Bengal, who in 1906 provided the detailed

---

[70] Henry Pearse, *The History of Lumsden's Horse; a Complete Record of the Corps from Its Formation to Its Disbandment* (London: Longmans, Green & Co, 1905), 410, www.archive.org/details/historyoflumsden00pearrich.

plan for the extension of Edward Henry's Pass Office fingerprint registration to the compulsory and universal registration of the Indians and the Chinese in the Transvaal. His elaborate plan was prompted by a terse suggestion from one of the detectives of Henry's CID that 'impressions of all ten fingers should be taken instead of the [unrolled] right hand thumb and that a Central Office for recording same should be instituted'.[71] Drawing on Galton's argument that the use of 'such common names as, say, Mohamed Ali' made it impossible to distinguish the claims of immigrants, Chamney derided the existing certificates – 'small square documents printed in triplicate on thin paper' that were 'utterly unsuited to last out the holder's lifetime'.

It was also Chamney who invoked the utilitarian principle that would later enrage Gandhi, arguing that 'the object to be kept in view was the greatest good for the greatest number' in justifying the subjection of the tiny Asian minority. He dismissed Gandhi's objections to full-print registration by suggesting that they were 'purely of the sentimental order' and, carefully blurring the boundary between single-print and ten-print registration, he argued that fingerprints were being routinely used in India 'as a means of identification'. (Interestingly, as we will see in the next chapter, when the Transvaal sought to register Indian immigrants, Edward Henry sided with Gandhi in protesting against the necessity for universal registration.)

It was this plan for universal ten-fingerprint registration that was championed by Milner's disciple, Lionel Curtis, and inherited by Smuts. The scheme extended the logic of Henry's fingerprint registries from the confines of the mining Labour District out into the general population and the whole of the Transvaal. But it also precipitated a battle over the compulsory fingerprint registration of black people that would provide a continuous vein of conflict, domestically and internationally, that came to define twentieth-century South Africa. Looking back, as the struggle with Gandhi was reaching its apex, Chamney's explanation for his plan was an orientalist paroxysm, a precisely reversed reflection of Gandhi's own image of the virtuous Indian 'who wants himself to be identified'. 'The Finger Print System was adopted in the Transvaal on my advice in the year 1907', he ranted in 1912, 'because after four years experience I had come to the conclusion that it was absolutely impossible to identify through their signatures or photographs this class

[71] See Annexure A in Chamney, Protector and Registrar of Asiatics to Lionel Curtis, Assistant Colonial Secretary (Division II), 'Report on the Position of Asiatics in the Transvaal (irrespective of Chinese Indentured Labour) in Relation Especially to Their Admission and Registration', 17 April 1906, TAD, LTG 97, 97/03/01 Asiatics Permits, 1902–7, TAD.

of Races who think nothing of changing their names and their history, the history of their Fathers and Mothers and everything else to further their own purpose'.[72]

Chamney's hysterical explanation seems a snug fit with our contemporary theories of the subversive qualities of unsanctioned histories. It is tempting to suggest that compulsory fingerprinting was triggered in the Transvaal by an epistemological conflict between British officials and Gujarati traders. But the truth was probably more banal.[73] Full print registration was imposed on the Indians of the Transvaal, and eventually on the entire population of the country, because the political arrangements in South Africa allowed it. White constituents, who were subjected to fingerprinting only upon arrest in the early years, bombarded their ministers with furious complaints about the absence of a 'wholesome discretion' in the police. These complaints prompted the Secretary of Justice to suggest that only those without 'fixed address known' to the police should be fingerprinted; the same officials endorsed the wholesale application of the control technology of imperial rule to all black people. As we will see in the next chapter Gandhi was an early advocate of this kind of government and eventually its most powerful critic.

## Conclusions

Edward Henry left South Africa from Cape Town on 24 April 1901, berthed in 'a very nice outside cabin' on the upper deck of the Union Castle *Carisbrooke*. He went to take up the position of Deputy-Commissioner of the Metropolitan Police, a post that he would hold for a little more than a year before ascending to the top post at Scotland Yard. For the next fifteen years he 'pulled the Met to the technological cutting edge'.[74] Henry's tenure in London also marks a break from the very thin forms of nineteenth-century policing in England to more invasive and secretive forms of surveillance that became the norm in the twentieth century. During this time the Metropolitan Special Branch, which had scarcely existed in the nineteenth century, grew to a force of 700 men, and began the routine surveillance of domestic politics. Much of the expansion of this police power, including the formation of MI5 and the establishment of the secret Register of Aliens, was hidden from parliamentary enquiry

[72] Chamney, Registrar of Asiatics, Department of the Interior to Acting Secretary for Justice, 'Confidential'.
[73] R. Guha, 'The Prose of Counter-Insurgency', in *Selected Subaltern Studies* (New York: Oxford University Press, 1988), 45–84; Ashis Nandy, 'History's Forgotten Doubles', *History and Theory* 34, no. 2 (May 1995): 44–66.
[74] Sengoopta, *Imprint of the Raj*, 183.

or the British press by the operations of the revised Official Secrets Act of 1911. There was, as Bernard Porter has observed, much in Edward Henry's term of office that seemed to give substance to William Morris' warning that the techniques of imperial government would find their way home.[75]

In fact the development of modern fingerprinting followed a very different set of pathways (notwithstanding the very large scholarship that now exists on the many forms of fingerprinting in the West). Henry's system of fingerprinting spread rapidly from the Witwatersrand as an essential part of the skeletal workings of the African gatekeeper state. It was taken by experts trained by the CID in Pretoria to the Congo Copperbelt in 1912, to Kenya in 1920 and 1930 for the making of Milner's Kipande system, to Sierra Leone in 1941, and then, again, to Kenya during the Mau Mau emergency.[76] In each case, Henry's system of fingerprinting equipped the colonial state with what might be described as the most basic tools of identification. Unlike bureaucracies elsewhere that emerged very largely from the obligations and responsibilities of local government, the colonial states in Africa were both incapable of and disinterested in building up the institutions of written civil registration. Fingerprinting well suited those limited capacities. It was precisely because it contradicted the long-established patterns of respectability that were associated with written civil registration, as Higgs' study of identification in England shows, that fingerprinting was not adopted in Britain.[77] And the fact that compulsory fingerprinting bore the historical taste of imperial subjection – long before the development of the new extensive body of privacy law – has helped to block its return to Europe.[78]

More intriguing is the fact that – until very recently – plans for compulsory fingerprinting did not return to India. The jettisoning of the Indian technology of fingerprinting in South Africa after 1910 can be explained in part, as we will see in the next chapter, by the global

---

[75] Bernard Porter, *The Origins of the Vigilant State: The London Metropolitan Police Special Branch before the First World War* (London: Boydell & Brewer Inc, 1991), 164–71; ironically, it was Milner who dismissed Henry after a police strike in 1918. Dilnot, *The Story of Scotland Yard*, 156; Higgs, *The Information State in England*, 109–11.

[76] *Request of Belgian Government to Be Supplied with Information Regarding the Finger Print System*, vol. 111, GG, 1912; *Transmits Request of Belgian Govt. to Be Supplied with Certain Information as to the Working of the Fingerprint System in South Africa*, vol. 109, GG, 1912; *British East Africa. Offer of Appointment as Finger Print Expert to Lieutenant Wcc Burgess Transmits Information Regarding Conditions of Service*, vol. 774, GG, 1919; *Visit of JD Grant from Kenya Colony for Instruction in Finger Prints at Criminal Record Bureau Pretoria*, vol. 1/2/278, PM, 1930; *Secondment of Fingerprint Experts to Kenya*, vol. 512, SAP, 1954; *Fingerprint Expert Loan of Services to Government of Sierra Leone*, vol. 347, SAP, 1945.

[77] This is the broader point of Higgs, *Identifying the English*.

[78] Higgs, 'Fingerprints and Citizenship'.

notoriety of Gandhi's conflict with the South African state. But there is more at work here than the caution of the British administration after 1920. For fingerprint identity registration was attempted on the Kolar Gold Fields in 1930, under circumstances that were clearly modelled on the arrangements on the Witwatersrand. As Nair shows, Indian workers resisted these plans with a ferocity that astonished their managers, because they – like their English peers – rejected fingerprinting because of its association with abjection. In India it was the history of the use of Henry's technology to isolate criminal populations and, especially, its use in the fortification of debt contracts that motivated workers to resist uncompromisingly. A century after Henry left India for South Africa that seems to be a history that is now forgotten.[79]

---

[79] Janaki Nair, 'Representing Labour in Old Mysore: Kolar Gold Fields Strike of 1930.' *Economic and Political Weekly* 25, no. 30 (July 28, 1990), 73–86. My thanks to Charles van Onselen for this reference.

3    Gandhi's biometric entanglement:
     fingerprints, satyagraha and the global
     politics of *Hind Swaraj*

Gandhi wrote *Hind Swaraj*, the anti-colonial manifesto that made him
one of the key political figures of the twentieth century, after six years
of struggle over the fingerprint registration of Indians in the Transvaal.
Many scholars have commented on the extraordinary change in his
politics in this period, and some have pointed to the special role that
the struggle with the Transvaal state played in the development of his
political philosophy.[1] But all of these studies rest upon a simplification
of Gandhi's role in these events that effectively clouds our understanding of the origins of satyagraha and the distinctively anti-progressive
politics he announced in *Hind Swaraj*. In this chapter I want to show
that Gandhi's entanglement in the design of the systems of identity in
South Africa in the first decade of the twentieth century was the source
of his discontent with progressivism. Before 1905 Gandhi shared a critique of modernity with those – like Milner and Curtis – who wielded
power in South Africa; he was a product of late nineteenth-century
anti-modernism, drawn to the same critiques of laissez-faire as the
imperial progressives. This transformation of anti-modernism in the
early decades of the twentieth century was a common process globally. Yet, unlike his peers in the United States and England (whom
Jackson Lears has followed coming to a meek accommodation with

---

Sections of this chapter appeared as 'Gandhi's Progressive Disillusionment: Thumbs,
Fingers, and the Rejection of Scientific Modernism in Hind Swaraj', *Public Culture* 23,
no. 2 (1 April 2011): 331–48.

[1]  Surendra Bhana and Goolam Vahed, *The Making of a Political Reformer: Gandhi in South
   Africa, 1893–1914* (New Delhi: Monahar, 2005), 19; Paul F. Power, 'Gandhi in South
   Africa', *The Journal of Modern African Studies* 7, no. 3 (1969): 450; Judith Margaret
   Brown, *Gandhi: Prisoner of Hope* (New Haven: Yale University Press, 1991), 55, 87;
   Uma Dhupelia-Mesthrie, *Gandhi's Prisoner: The Life of Gandhi's Son Manilala* (Cape
   Town: Kwela Books, 2007), 76–7; Louis Fischer, *The Life of Mahatma Gandhi* (New
   York: HarperCollins Publishers Ltd, 1997), 109; M. Prozesky and J.M. Brown, *Gandhi
   and South Africa: Principles and Politics* (Pietermaritzburg: University of Natal Press,
   1996); M. Swan, *Gandhi: The South African Experience* (Johannesburg: Ravan Press,
   1985), 163.

corporate managerialism[2]) Gandhi broke decisively, and irrevocably, with progressivism in August 1908. This rupture was prompted by his realisation that the progressives were wedded to the ideology of segregation – and the use of the technologies of fingerprinting – in order to carve out a racially defined state. As I show here, the ferocity of Gandhi's rejection of managerial progressivism was driven by the bitter contradictions between his protests and his participation in the design of this segregationist order.

The protests that Gandhi led against fingerprinting in 1906 and 1907, and after a period of acquiescence, again after August 1908, were organised around the danger the practice represented to the respectability of Indian women and children, and to the honour of Indian men, in South Africa and elsewhere. The 'gendered discourse of national honour' that Mongia has recently traced in the successful mobilisation of satyagraha to defend Indian marriages in South Africa in 1913 has roots in the earlier fingerprinting campaign. Gandhi's effort to redefine honour around the decontextualised issues of compulsion and dignity after the compromise with Smuts in January 1908 flew in the face of this gendered, and deeply emotional, doctrine, setting him at odds with the constituency he had formed and mobilised.

If the political wreckage after his effort to reverse the direction of gendered outrage was one key to understanding Gandhi's bitter rejection of Western modernity in general, and colonial government in particular, in the *Hind Swaraj*, another was his very practical involvement in the design of the administrative procedures of progressive imperialism in the Transvaal. Contrary to the popular view of his role, Gandhi saw himself as an expert administrator and an architect of more efficient and secure legal mechanisms for regulating the movement and identity of Indians in South Africa.[3] He was a precocious and successful advocate of administrative fingerprinting for South African Indians. In 1904 when he recommended to the Natal government that illiterate Indians should be required to provide thumb-prints on promissory notes the law was revised to accommodate his suggestion. While he was in the thick of the conflict with Smuts he became an enthusiast of Edward Henry's *Classification and Uses of Finger Prints*, and an expert on the administrative costs and benefits of ten-print and thumb-print registration. When he endorsed full-print registration in 1908 he was accepting Smuts' argument that

---

[2] Jackson Lears, *No Place of Grace: Antimodernism and the Transformation of American Culture, 1880–1920* (New York: Pantheon, 1981).
[3] In this respect McKeown is correct that Gandhi's politics encouraged the bureaucratisation of immigration systems, see Adam McKeown, *Melancholy Order: Asian Migration and the Globalization of Borders* (New York: Columbia University Press, 2008), 317.

the state required a scientific basis for identification, and he used the same scientific virtues of ten-print registration to cajole the Indians in the Transvaal to register.

The emphasis Gandhi placed on consent worked to bolster the reasonableness of his protests against stigmatising administrative requirements, and, after 1908, he argued that it provided a tool to dissolve the hold of the administrative procedures he had helped design. But in this respect he was wrong. Despite his claims, at the time and many years afterwards, the Asiatic fingerprint registry that was built in these months remained a tool of policing, taxation and movement control of precision and longevity unprecedented in the history of the South African state. Despite his constant claims to the contrary afterwards there was no way to withdraw the fingerprint registrations once they had been offered. It was substantially for this reason that in the year before the writing of *Hind Swaraj* he was abandoned and castigated by the group he had toiled to represent. His precocious and furious rejection of what Nandy has called technologism in 1909 makes particular sense in the light of this horrible predicament.[4]

Much of what Gandhi believed in 1909 about the virtues of traditional India he learned from *fin de siècle* anti-modernism.[5] He rediscovered the virtues of the *Gita* in London after the encouragement of two English theosophists.[6] Two of the three key intellectual figures in his life, Ruskin and Tolstoy, were the pillars of middle-class criticism of industrial society in the Atlantic. The appendix of literary authorities that he appended to the *Hind Swaraj* reads like a short list of recommended readings for nineteenth-century anti-modernism, stressing his devotion to Tolstoy, Carpenter, Ruskin and Thoreau. After working in the cities of Durban and Johannesburg for almost his entire adult life, even Gandhi's account of the political virtues of the Indian village was derived from Henry Sumner Maine's *Village Communities in the East and West*;[7] his preoccupation with handspinning as the remedy for the economic ills of the sub-continent came from Birdwood's *Industrial Arts of India*.[8] I think it

---

[4] Ashis Nandy, 'From Outside the Imperium: Gandhi's Cultural Critique of the "West"', *Alternatives* 7 (1981): 176–7.

[5] See Nandy, 'From Outside the Imperium'.

[6] Brown, *Gandhi*, 25.

[7] M.K. Gandhi, *Hind Swaraj and Other Writings*, ed. A.J. Parel (Cambridge University Press, 1997), xlii–xlv; Lears, *No Place of Grace*, 75–82; see, on Maine, M.K. Gandhi, 'Baroda: A Model Indian State', *Indian Opinion*, 3 June 1905, *Collected Works*, vol. 4, 302.

[8] That Gandhi had read Birdwood and Maine is clear from M.K. Gandhi, 'Petition to Natal Legislative Assembly', 28 June 1894, *Collected Works*, vol. 1; Patrick Brantlinger, 'A Postindustrial Prelude to Postcolonialism: John Ruskin, William Morris, and Gandhism', *Critical Inquiry* 22, no. 3 (Spring 1996): 466–85.

is indeed significant that, as Nandy pointed out, 'almost all of Gandhi's gurus were Western intellectuals'.[9] The significance is double-sided – explaining much, on the one hand, about Gandhi's politics, but also, on the other, about the global history and politics of anti-modernism viewed outside of the national frames of American (or British) history. The fingerprint registration struggle in the Transvaal was 'the crucial moment for Gandhi and the history of colonial resistance movements', not because it provided a new kind of political struggle in satyagraha, but because, through the *Hind Swaraj*, it reinvigorated and redirected nineteenth-century anti-modernism around the world.

At precisely the same time as the critics of industrialism in the United States and England were gradually making peace with the expert-led factory system, Gandhi was being pushed away from it. In the years between 1893 and 1908 Gandhi offered a continuous stream of helpful suggestions to the colonial government to iron out the kinks, loopholes and unnecessary injustices of colonial administration. In the manner in which he ran his legal office, and in his relationships with officials in Natal and the Transvaal, he functioned in this period as a volunteer managerialist bureaucrat. That practice came, mostly, to an end in 1908. His turn to Tolstoy's simple life and Ruskin's critique of industrial cap-italism came after the struggle over the fingerprint registration of the Transvaal Indians had begun in earnest. The 'remarkable transformation in Gandhi between 1906 and 1909' that many scholars have discerned was his movement from managerialism to anti-modernism, precisely the opposite journey to the one undertaken by key progressive critics of industrialism like Beatrice Webb and Jane Addams.[10]

### Thumb-printing

Gandhi arrived in Natal in 1893 to represent one of the largest Gujarati firms in a dispute with another. Very quickly he helped establish the Natal Indian Congress, and found himself caught up in the merchants' strug-gles against the Natal Colony's efforts to deny Indians basic rights of citi-zenship and the Boer Republics' plans to expel them from the towns. In 1896, during what turned out to be a short trip back to India, he displayed

[9] Nandy, 'From Outside the Imperium', 172.
[10] Bhana and Vahed, *Gandhi in South Africa*, 19; Lears, *No Place of Grace*, 81–3; see also Power, 'Gandhi in South Africa', 450; Beatrice Potter Webb, *'Glitter Around and Darkness Within,' 1873–1892*, ed. Jeanne MacKenzie and Norman Ian MacKenzie (Cambridge, MA: Belknap Press of Harvard University Press, 1982); Beatrice Potter Webb, *All the Good Things of Life, 1892–1905*, ed. Norman Ian MacKenzie and Jeanne MacKenzie (Cambridge, MA: Harvard University Press, 1983).

his remarkable skills as a newspaper propagandist with the publication of the Green Pamphlet on the 'Grievances of the British Indians in South Africa'.[11] The pamphlet skilfully exploited the telegraphic reporting links between papers, reaching a global audience, and so outraging the white settlers back in Natal that he was very lucky to avoid lynching when he returned to Durban at the end of the year. In the short time that followed before the South African War he assisted the British to build the case against the Republics in part for their treatment of the Indians. He famously served as a stretcher bearer for the British troops in the disastrous early stages of the conflict, before, again, returning to India.[12]

When the Natal Indian Congress brought Gandhi back to South Africa early in 1903, he arrived with the expectation that 'our position in the Transvaal is and ought to be infinitely stronger than elsewhere'.[13] He quickly began to realise that the British government under Lord Alfred Milner had plans to exercise a more onerous set of identification controls over the mixed population they defined as Asiatic. When Gandhi managed to secure an audience with Milner, in June 1903, the organisations representing the interests of Indian merchants protested the new administration's plans for enforcement of the old Republic's Law 3 of 1885. The main elements of this newly enforced law combined the worst of the discriminatory practices from Natal and the Boer Republics, including draconian limits on Asian immigration, a registration fee of £3 for every adult male, and the threat of the restriction of their trade to segregated bazaars outside of the town centres.

A politics of the technology of identification registration quickly moved to the foreground of this conflict. At the meeting with Milner, Gandhi protested that Indians were being forced to provide three photographs in order to secure passes to leave, and return, to the colony. He objected that this special requirement implied that 'all Indians were criminally inclined', but, as he remarked repeatedly over the next five years, he also found the use of photographs invasive, and abusive, bearing the taint of criminality.[14] Milner promised to consider 'the points you have made about photographs, about the difficulty of getting the title to

---

[11] Saumendranath Bera, 'Confronting the Colonial Communication Order: Gandhiji, the Green Pamphlet and Reuter', in *Webs of History: Information, Communication, and Technology from Early to Post-Colonial India*, ed. Amiya Kumar Baghi, Barnita Bagchi and Dipankar Sinha (New Delhi: Indian History Congress, 2005), 209.

[12] Swan, *Gandhi*, 53–88; Bhana and Vahed, *Gandhi in South Africa*.

[13] Swan, *Gandhi*, 58.

[14] M.K. Gandhi, 'The British Indian Association and Lord Milner', *Indian Opinion*, 11 June 1903, *Collected Works*, vol. 3, 64–73; M.K. Gandhi, 'Oppression in the Cape', *Indian Opinion*, 29 December 1906, *Collected Works*, vol. 6, 201; M.K. Gandhi, 'A Dialogue on the Compromise', *Indian Opinion*, 8 February 1908, *Collected Works*, vol. 8, 136–47.

mosques registered in your own names, and about passes' but he also announced that the state intended to create a special purpose Asiatic Department, and – adopting Henry's recommendations – to impose a systemic programme of identity registration on Indians particularly. One deeply significant result of this cordial meeting, and the negotiations for the simplification of the permit system that followed, was that the representatives of the Indians in the Transvaal agreed to take out a new set of registration certificates, incorporating, for the first time, the use of the thumb-print as a marker of identification.[15] These documents, which Gandhi later claimed were adopted voluntarily 'to please Lord Milner', marked the beginnings of a system of identity for Indians in South Africa that hinged on fingerprinting.[16]

Gandhi's own views on thumb-prints at this time were shaped by his efforts to foster a politics of discrimination, to separate out his wealthy, literate clients from the broader mass of indentured and ex-indentured workers. A year after his meeting with Milner, Gandhi wrote to the Attorney General of the Colony of Natal urging him to include a legal requirement that illiterate Indians should be required to provide a thumb-print on any contracts of debt. 'I venture to think that if in this excellent measure a clause', he noted of the draft bill regulating debt contracts for Indians, 'is inserted that those who cannot sign their names in English characters should, in addition to putting the mark, put their thumb impression also, it would be a complete measure for the safeguard sought for in the bill'. While the officials in Natal were calling for 'promissory notes not written and signed in English in the makers own hand-writing' to be invalid unless made in the presence of a magistrate, Gandhi suggested, with his experience of the Transvaal permit system in mind, only a 'thumb impression would completely protect innocent persons'.[17]

A week later he elaborated on the same point in *Indian Opinion*. After first congratulating the Natal government for introducing the bill to control the signing of promissory notes by Indians, he worried about the great legal weight that would be accorded to notes signed before a government representative. 'It has been found that it is impossible to forge a thumb-mark', he advised his readers, 'and the thumb-impression would be the

[15] M.K. Gandhi, 'Fair and Just Treatment', *Indian Opinion*, 11 August 1906, *Collected Works*, vol. 5, 300.

[16] M.K. Gandhi, 'Statement by the Delegates on Behalf of the British Indians in the Transvaal Regarding the "Petition" from Dr. William Godfrey and Another and Other Matters', 20 November 1906, *Collected Works*, vol. 6, 126.

[17] M.K. Gandhi to Attorney General, Natal, 'Letter to Attorney General', 30 June 1904, MJPW 128 in MJ1287 'MK Gandhi Re bill to regulate the signing of negotiable instruments by Indians. Suggests the finger prints also be taken', 30 June 1904. Magistrates' suggestions are in the same file.

surest safeguard against impersonation, for it may happen that the man who may put his mark before a Magistrate or a Justice of the Peace may not at all be the person intended to be charged with the debt'.[18] Gandhi's daily negotiations for his clients suggest that he viewed the thumb-print as a reliable administrative remedy to the 'question of fraud'. As late as August 1905 his application for a certificate of return for Abdul Kadir, one of the wealthiest men in Durban and the president of the Natal Indian Congress, noted approvingly that 'the thumb impression on the certificate you may issue would prevent its use by any one else'.[19]

There is no sign in these early interventions of Gandhi's later concern with the implications of what he, and others, called class legislation or the racial taint – the special, criminalising focus of legislation on Indians as a group. Indeed, his role in the years between 1902 and 1906 was to act as an expert – and often as an advocate – of special legislation targeting Indians. This is probably because he did not see any reason to articulate the need for exemptions for 'well-known' members of the merchant elite. His position changed early in 1906 as the state, chiefly through Lionel Curtis' office, began to plan to build a ten-print fingerprint register and to apply the routines of thumb-print identification where impersonation was unthinkable.[20] When the permit officer at the border post at Volksrust on the road from Durban to Johannesburg 'had the effrontery to ask Mr Johari', the representative of the Natal firm of Aboobaker Amod & Bros – 'a cultured Indian' who had 'travelled in Europe and America' – to 'put his thumb-impress on his book', Gandhi protested the unmistakable racist insult. Thumb-prints – at least for respectable, known and literate members of the Indian elite – were horribly degrading. 'Well may Mr Johari ask', he wrote to *Indian Opinion*, 'whether he is to be treated as a criminal, without being guilty of any offence, save that of wearing a brown skin'.[21]

## Scientific thumbs

From the start of 1906, Lionel Curtis began to make the case for an elaborate, centralised fingerprint registration scheme designed as the lynchpin of a segregated state, to 'shut the gate against the influx of an Asiatic

---

[18] M.K. Gandhi, 'Indian Promissory Notes', *Indian Opinion*, 2 July 1904, *Collected Works*, vol. 4, 28.

[19] Interestingly Gandhi makes no mention of the history of thumb-printing in India in these recommendations. See, for that history, Chandak Sengoopta, *Imprint of the Raj: How Fingerprinting Was Born in Colonial India* (London: Macmillan, 2003).

[20] M.K. Gandhi, 'Deputation to Colonial Secretary', *Indian Opinion*, 17 March 1906.

[21] M.K. Gandhi, 'A Contrast', *Indian Opinion*, 10 March 1906, *Collected Works*, vol. 5, 113.

population' and to 'guard the Transvaal as a white reserve'.[22] For the next two years Gandhi campaigned locally and internationally to limit the effects of this register without contradicting the 'principle of white predominance'. In this period Gandhi, and his allies, had worked assiduously to represent ten-fingerprint registration as a mortal threat to the honour of Indians, in South Africa and globally.

The government's plans for a new system of registration in the last months of imperial rule were announced in an Asiatic Ordinance by Patrick Duncan, the Colonial Secretary, at the beginning of August 1906. The promise of a new round of registration, with unspecified but ominous requirements for identification, made Gandhi cancel a planned trip to London to lobby the Colonial Office on the wider constitutional process. When the draft bill was actually published on the 22nd of that same month it prompted immediate protests against the requirements for another round of registration. There was no mention of fingerprinting in the bill itself. The first round of these protests all commented on the injustice of the government answering 'a plea for relief' with even more harsh registration, but the most explosive protests dealt with the position of women and children under the new regime.

Running through all the responses to the ordinance was the deeply emotive claim that the law represented a threat to 'female modesty, as it is understood by millions of British Indians' and that it would 'ride roughshod over sentiments cherished dearly for ages'.[23] The cable that the British Indian Association dispatched to the Indian press read like a tabloid headline. The law it announced 'shocks Indian sentiment by requiring women, and children over eight years to register'. Indians in the Transvaal preferred the old law of the Boer Republic to the 'wanton indignity which the proposed ordinance contemplates'. The articles published in Gujarati in *Indian Opinion* continued on the same theme. Under the headline 'Abominable' Gandhi observed that the Asiatic Act 'unsettles the Indian mind as no other measure in South Africa has ever done before. It threatens to invade the sanctity of home life'. In the same edition, under the headline 'Criminal', he asked whether women and children will 'be banished from the Colony and torn from their husbands, or parents, as the case may be?'[24]

[22] Lionel Curtis, Assistant Colonial Secretary (Division II) to Patrick Duncan, Colonial Secretary, 'Position of Asiatics in the Transvaal', 1 May 1906, LTG 97, 97/03/01 Asiatics Permits, 1902–7, TAB.
[23] Abdul Gani, Chairman, British Indian Association to Patrick Duncan, 'Letter to Colonial Secretary', 25 August 1906.
[24] M.K. Gandhi, 'CABLE TO "INDIA"', 28 August 1906; M.K. Gandhi, 'CRIMINAL', *Indian Opinion*, 8 September 1906; M.K. Gandhi, 'ABOMINABLE', *Indian Opinion*, 8 September 1906.

After puzzling over the elite concerns and protests of the British Indian Association in the years leading up to the Black Act, Swan has looked for explanations as to why '3000 people were suddenly mobilised for passive resistance in September 1906?' Part of the answer lies in these gendered protests. Like the public discourse of satyagraha in 1913, which Mongia observed was 'a defense of the honour of Indian women … coterminous with the honour of the Indian nation', Gandhi's assault on what would become the 1907 Black Act mobilised an intense and universal sentimentality about gender roles.[25] The articles announcing the 'Criminal' and 'Abominable' dangers to Indian families appeared in *Indian Opinion* just three days before the famous mass meeting that launched the campaign of peaceful resistance on 11 September 1906. If Swan is correct that the Asiatic Ordinance was a catalyst that Gandhi used to move his audience towards the self-sacrificing convictions of satyagraha then it is also true that his protests had a momentum and direction of their own. By couching his moral lessons in the language of gendered honour Gandhi left open only a small possibility for the kind of dramatic political reversals he would later advocate.

When the 3,000 people gathered inside the Empire Theatre on Commissioner Street in the afternoon of Tuesday, 11 September 1906, the government had already conceded that women would not be subjected to registration. Lionel Curtis' decision to demand ten fingerprints from the Indians had not yet been officially announced, but the leaders of the protests certainly had forewarning (from their meetings in March) of what he had in mind. From this point the intimate machinery of identification moved to the foreground of the protests against the Asiatic Ordinance. In front of Chamney, who was sitting on the theatre stage, Nanalal Shah protested against the requirement for new registration. 'This register contains my name, my wife's name, my caste, my profession, my height, my age', he protested, holding up his Crown Colony registration certificate. 'It bears even my thumb-impression. Is all this not enough? How can anyone else use this register? Does the Government want now to brand us on our foreheads?' Speaker after speaker followed him promising to go to jail before they would undergo another round of registration.[26]

Some time between the Empire Theatre meeting and Gandhi's departure by steamship for London on 1 October, Lionel Curtis met with the leaders of the British Indian Association and confirmed that his plans

[25] Radhika Mongia, 'Gender and the Historiography of Gandhian Satyagraha in South Africa', *Gender & History* 18, no. 1 (2006): 130–49.
[26] Gandhi, 'Johannesburg Letter'.

included a 'system of identification under which all Indians would be required to put down on their passes the impressions of their ten fingers'.[27] The fact that applicants for the new identification certificates had to provide a full set of fingerprints was not the only objection. The initial application was combined with the requirement in the ordinance that Indians would be compelled not only to carry passes but to provide the 'means of identification as may be prescribed by the regulations' to police officers on the street 'which, according to Mr. Curtis's declaration, means finger-impressions'. This new fingerprint identification was, as Gandhi wrote to *The Times*, 'a system of passes and identification applicable only to criminals'.[28]

Patrick Duncan's rapid capitulation on the requirement that Indian women would also be subjected to the new methods of identification did little to persuade Gandhi to abandon the discourse on the dangers to the Indian family. Immediately before he left for London he picked up on the story of a woman named Punia who had been forced off the train, with her husband, at the Volksrust border. At the trial that followed, Gandhi reported, the policeman testified that he had orders to arrest 'permitless women and children whether or not they were accompanied by their husbands or parents and whatever the age of the children'. In the same edition, *Indian Opinion* commented that 'Mr. Gandhi has understated the gravity of the situation, in that he omitted to mention what is, perhaps, the most unpleasant feature of an unfortunate affair, namely, that the woman was made to give her ten finger-prints at the Charge Office in Volksrust, and was obliged to do so again at Germiston'. Gandhi skilfully interwove the specific case of arrest, fingerprinting and the wider derogatory comments from the press and officials into an outraged defence of the 'infamous lie' that Indian wives were 'often of indifferent character'.

In the same edition (all of this in the week prior to his departure for England) Gandhi drew out the implications of the conviction of children for violations of the permit regulations. The police had used a thumb-print taken from a ten-year-old boy, Mohamed, travelling into the Transvaal from Natal with his father, Hafeji Moosa, to argue that both father and son were 'guilty of obtaining a permit by improper means' because the details on the permit did not match the identity of the child. The impressions taken at the border were 'sent to Pretoria, and as they did not tally with the thumb-imprints on the counterfoil' man and child were later arrested in Potchefstroom. The proceedings of the court hearing

---

[27] M.K. Gandhi, 'Interview to "South Africa"', *Indian Opinion*, 1 November 1906.
[28] M.K. Gandhi and H.O. Ally to Editor, *The Times*, 'Letter to "The Times"', 22 October 1906.

highlighted exactly the kinds of familial danger that Gandhi was warning about: 'The Magistrate discharged the father, but found the [ten-year-old] son guilty, and sentenced him to pay a fine of £50 or suffer imprisonment without hard labour for three months.' All of these reports belaboured the same point – 'thus are wives separated from husbands and children from parents under this Government' – and the same remedy: 'it is a thousand times better for men to suffer imprisonment than to submit to such a law.' [29] In his struggle against the 1906 Ordinance, and again later, against Smuts' Act 2 of 1907, Gandhi would repeatedly return to the dangers that fingerprinting presented to the imperilled Indian family. [30]

Like the press reports, the written deposition that Gandhi presented to Lord Elgin, the Colonial Secretary in the new Liberal government, turned on the criminalising humiliations of compulsory fingerprinting and the attendant administrative violation of the Indian family. Curtis' fingerprinting design imposed a regime of identification on the British Indians in the Transvaal that had only ever been 'applied to the worst criminals'. The most moving part of the deposition was a detailed list focused on the danger that the system represented to children, who were required to register provisionally in order to secure their rights of residence in the Transvaal: 'A baby eight days old will have to give ten digit prints and be carried to the registering officer.' [31]

As a former Indian viceroy, Elgin may have been especially susceptible to the global protests of humiliation that Gandhi brought with him to London. The South African delegation was certainly accompanied to the Secretary of State's office by a formidable array of Raj luminaries and parliamentary representatives. In his introductory speech Sir Lepel Griffin, president of the East India Association, recapitulated the protests that had emerged from Johannesburg over the previous two months. 'Under this regulation', he protested, 'every Indian in the Transvaal, whether an adult male, whether a woman, or whether a child, and even babes in arms will be obliged to be registered under such conditions as ordinarily apply only to convicts in a civilized country'. The criminalisation of the Transvaal Indians was an affront not just to the '300 millions of Indians' but also to the 'whole body of Indian officials to which I and most of the members of this deputation belong, who are insulted along with the natives of India'.

---

[29] M.K. Gandhi, 'The Punia Case', *Indian Opinion*, 29 September 1906.

[30] M.K. Gandhi, 'Deputation to Lord Elgin', 8 November 1906; M.K. Gandhi, 'Johannesburg Letter', *Indian Opinion*, 21 September 1907; M.K. Gandhi, 'Letter to "The Rand Daily Mail"', *Indian Opinion*, 9 October 1907.

[31] M.K. Gandhi and H.O. Ally, 'Representation to Lord Elgin', *Indian Opinion*, 31 October 1906.

In the presentation that followed Gandhi was quick to point out that the Ordinance no longer applied to women, but he strongly endorsed Griffin's presentation. He also reminded Elgin of the Punia case – 'a poor woman torn away from her husband' – and the eleven-year-old boy 'sentenced to pay a fine of £50 or go to gaol for three months, and at the end of it to leave the country'. And then, in order to make the point about the invasiveness of the Transvaal administration, he showed the Colonial Secretary an old Republican registration certificate – a simple receipt for payment of £3 – and his own Crown Colony certificate in order to show 'how complete it is to establish identification'. That was a mistake, because it created an opportunity for Lord Elgin, the old India hand, to philosophise on the politics of 'this question of thumb marks'.

Elgin was viceroy between 1894 and 1899, exactly the period that Edward Henry had been energetically introducing fingerprinting, and thumb-printing, into the administration of the government of India. By the end of his reign, thumb and finger prints were being used to control identity on pensions, opium contracts, municipal job applications, money orders and even examinations.[32] For Elgin, obviously deeply skilled in the dark arts of deflecting petitioners, this proliferation of digit impressions could be seen as 'a marvelous thing'. Pointing to Gandhi's registration certificate his closing comment was clearly intended to finesse the outrage of his petitioners. 'I want just to mention, and to bring to the notice of Mr. Gandhi, that on the permit which he has handed to me, issued under the present Ordinance', the ninth Earl observed, 'there is a thumb mark already imposed under the present Ordinance in just the same way as it will be imposed under the new Ordinance'.

Trapped by the rules of imperial etiquette that forbade him from speaking after the Colonial Secretary, Gandhi and the others struggled to articulate the distinction between fingers and thumbs in two or three staccato objections. The thumb impression was a 'purely voluntary act', which they had agreed to do because Milner had 'asked us to do it'. The debasement, Gandhi interjected, stemmed from the 'ten-finger mark'. By the end of the meeting he could only breathlessly appeal for a full commission of enquiry which would allow us to 'place our position accurately before your Lordship'.

Elgin concluded his comments by suggesting that the South African efforts to control Indian immigration formed part of a movement 'all over the world on the part of white communities, and we have to reckon with them'. And that, although he did not announce it, was his solution. Elgin

---

[32] Sengoopta, *Imprint of the Raj*, 150–3.

let the white leaders in South Africa know that he was going to withhold royal assent from the Ordinance, but that he would not do the same for an identical act passed by the new local government.[33] The bill for the Asiatic Law Amendment Act (No. 2 of 1907) – which Gandhi would call the Black Act – was the first item of government business, after the budget, in the newly elected parliament of the Transvaal Colony.

### Humbled will your manhood be

It was a sign of the shared earnest with which imperial officials, like Curtis, and the newly elected colonial leaders, like Smuts, approached the question of Indian immigration, that the first task of the new legislature was the passing of the Asiatic Law. The leaders of the British Indians had been preparing to protest the reading of the act for months when the Transvaal parliament opened on 21 March 1907, but they were astonished by the speed at which the law was rushed through committee and pushed through the assembly: 'None of us had any suspicion at the time', Gandhi told the protesters at the Gaiety Theatre nearly a week after the measure had already been passed, 'that the same law would be re-enacted within 24 hours and that normal parliamentary procedure would be suspended for the purpose'. This protest meeting marked the launching of a very effective alliance of merchants, hawkers and professionals resisting the terms of the act over the next six months. By the end of the year, when the first groups began to court imprisonment, even Smuts admitted that Gandhi had 'made a very successful resistance to the finger-print registration'.[34]

In the work to mobilise this alliance Gandhi described a barrage of horrible consequences that would follow from acceptance of the Asiatic Law. At the start of the campaign the most powerful of these was the threat of the quickening tempo of segregation, with its likely devastating economic effects for all Indians in the Transvaal. 'There is already talk today of relegating all the traders to Bazaars', he warned, 'of sending people to Klipspruit thirteen miles away from the Malay Location, and entirely abolishing the right of Indians to own land outside a Location'. Here he clearly articulated the consequences of meek compliance with the new

[33] W.K. Hancock, *Smuts: 1 the Sanguine Years, 1870–1919* (Cambridge University Press, 1962), 330. For the tightly connected struggles of white political leaders across the globe at this time, see M. Lake and H. Reynolds, *Drawing the Global Colour Line: White Men's Countries and the International Challenge of Racial Equality* (Cambridge University Press, 2008).

[34] Smuts, J.C. to Merriman, J.X., 8 January 1908, Selections from the Smuts Papers; see also Swan, *Gandhi*, 160.

round of registration. If the Indians targeted by the new Act failed to stick to their resolutions at the Empire Theatre meeting in September 1906 'on going to gaol', he warned grimly, 'they will lose everything'.[35]

A secondary theme of national humiliation also ran through his protests. The requirement of ten-print registration and pass-bearing would subject the Indians of the Transvaal to a criminal stigma, placing them beneath Africans and Malays in the local racial status hierarchy. Gandhi repeatedly stressed that it 'would be cowardice and betrayal of our country for us to endure it all silently'. The India that he had in mind here was an equal and constituent member of the imperial system, a junior but promising disciple in a meritocratic hierarchy. Drawing on the attractions of metropolitan acclamation he warned that the Transvaal Indians would bear responsibility for the fingerprinting of eminent imperial figures like the Tory MP for Bethnal Green, Sir Muncherji Bhownaggree. The Transvaal Indians would become the butt of jokes from the white press, and subject to an intolerable new burden of administrative nuisance. This last danger was especially obnoxious because it targeted the respectable and law-abiding – those who had already voluntarily subjected themselves to the colonial laws. 'It is not the people with forged permits or without permits who will have to face this harassment, since they will have left the Transvaal. It is the others with valid permits who will have to go through it all.'

Faced with a unanimously racist local parliament Gandhi began to argue that the most effective strategy for removing the racial taint was a combination of collective resistance to the unjust law and the offer of voluntary submission to the requirements of the Asiatic Registrar.[36] Towards the end of 1907, before the famous agreement with Smuts, he began to urge his audience to consent to a voluntary round of registration. To do this Gandhi insisted on the scientific qualities of thumb-print identification using his reading of Henry's *Classification and Uses of Finger Prints*. His first object was to persuade the audience of *Indian Opinion* that there was nothing degrading, or peculiar, about having to give up their thumb-prints. 'In England they have become a rage', he assured his readers, 'friends send their thumb-impressions to one another'. He reminded them of the widespread use of thumb-prints in India for government transactions, and of the fact that in Natal – he did not say that it was owing to his own recommendation – 'it is the practice to have thumb-impressions on promissory notes'.[37]

---

[35] M.K. Gandhi, 'Transvaal Asiatic Ordinance', *Indian Opinion*, 30 March 1907.
[36] Swan, *Gandhi*, 141–61.
[37] M.K. Gandhi, 'Johannesburg Letter', *Indian Opinion*, 28 December 1907.

Writing in Gujarati, Gandhi then explained the workings of the fin-gerprint repository. He contrasted the system of thumb-prints, which he saw as an aid to identification, with the registration of ten fingerprints directed at people who want to hide their identities. His account closely followed the explanation that Henry presented in his book. Fingerprints were taken from criminals precisely because they want to hide, or lie about their names, and the prints allowed the state to determine their identity. 'A person who has been required to give impressions of all fin-gers and thumbs can be identified by means of these impressions', he explained; using the different categories of patterns on the prints made, it was 'possible to prepare an index with the help of the impressions'. And it was the index that made it possible to identify someone, 'even if he has not given his correct name'.[38]

In the light of his stress on the special regulatory stigma directed at Indians, and the acute dangers of gendered dishonour, it is easy to see why his audience had trouble understanding his proposal to undertake a further round of voluntary registration.[39] For Gandhi, despite his prot-estation of the criminalising effects of ten-print registration, the issue was not 'merely one of the thumb or the fingers'. From the beginning he strained to make the argument that 'anything voluntarily accepted by us cannot be regarded as humiliation'. This stress on consent would later become one of the foundation stones of satyagraha; as Gandhi argued from India in the 1920s, 'the Government cannot exercise control over us without our cooperation'.[40] In the Transvaal in 1907 his views were more narrowly conceived around the virtues of consent for an imperial audience. He tried to persuade his audience that the doubtful success of his mission to Lord Elgin was owed to their earlier offer to under-take voluntary registration to satisfy Milner. Consent here was a tool for mobilising the support and sympathy of the liberal empire: 'our humility, forbearance and good sense will be appreciated in England.'[41]

Gandhi was insistent that, because the Indian traders in the Transvaal wanted to be recognised, ten-print registration was unnecessarily degrad-ing, and wasteful. He had no sympathy for the complaints of confusion and impersonation that issued from Chamney in the Asiatic Registrar's office, because he believed in the efficacy of the thumb-printing system;

---

[38] *Ibid.*
[39] Even the delegate who supported Gandhi's proposal admitted that he did not 'under-stand it fully'. M.K. Gandhi, 'Mass Meeting of Transvaal Indians: Full Account', *Indian Opinion*, 6 April 1907.
[40] Gandhi, *Satyagraha in South Africa*, vol. 3, *The Selected Works of Mahatma Gandhi* (Ahmedabad: Navajivan Publishing House, 1928), 218.
[41] Gandhi, 'Mass Meeting of Transvaal Indians: Full Account'.

he argued in court and in the newspapers that thumb-printing had successfully eradicated 'trafficking in permits'.[42] Fingerprinting, he argued, was completely unnecessary for the Indian in the Transvaal because he 'wants himself to be identified'. The special relationship between the Transvaal Indians and the Asiatic Register was at the core of this desire to be recognised. 'If his name is not on the records of the Government, he cannot live here', Gandhi explained; 'If he does not describe himself correctly, he cannot live in this country'. Running through this campaign in favour of thumb-prints, significant in the light of Gandhi's later discontent with science in *Hind Swaraj*, was his insistence that his 'argument has a scientific basis'.[43]

### Scientific fingers

As the time approached for the Asiatic Bill to become law, Jan Smuts, serving as the Colonial Secretary of the new responsible government in the Transvaal, began to have doubts about the wisdom of Curtis' elaborate plan to extract ten fingerprints from each of the Asians in the Transvaal. In June 1907 he wrote, as a 'matter of extreme urgency', to the state's legal advisers, and to the head of the Transvaal Criminal Investigation Division that Edward Henry had set up in 1900. After pointing out that the ten-impression requirement was derived from the system of Chinese indenture, he asked whether 'in view of the strong agitation on the subject it might be desirable to adhere to the present system of identification by which the imprint of the right hand thumb only is required'. But Smuts' question contained a rider, one which would provide the basis not only for his own insistence on ten-print registration, but also a dramatic reversal of Gandhi's own attitude. He was content to retain the thumbprint system 'provided always that the single imprint can be relied upon as sufficient evidence of identification to satisfy the courts'.[44]

It is important to keep in mind that in the decade after Edward Henry's departure from South Africa the legal basis of single fingerprint – what would later be called latent print – identification did not even exist in theory. There was, at this time, no meaningful statistical basis for the claims of the uniqueness of each fingerprint that had been popularised by Galton and Henry. (It was only in 1910 that the Parisian criminologist, Victor

---

[42] M.K. Gandhi, 'Trial of PK Naidoo and Others', *Indian Opinion*, 28 December 1907, *Collected Works*, vol. 8, 41.

[43] Gandhi, 'Johannesburg Letter', 28 December 1907.

[44] Assistant Colonial Secretary to Secretary to the Law Department, 'Letter to Secretary to the Law Department', 19 June 1907, LD 1466 AG2497/07 Finger Impressions, 1907, TAD.

Balthazard, elaborated on Galton's casual claims of the improbability of matching fingerprint minutiae to make the kind of statistical case that might support forensic uniqueness.[45]) In the meantime the claims that the new fingerprint experts were making about single-print identification were being received sceptically in the courts and outside of them.[46] In the face of this uncertainty the South African state's legal advisers opted for statistical certainty. 'The only really safe course', they argued, 'is to adopt a system of taking ten digit impressions'.[47] Here then was the argument that Smuts, the lawyer, would use to persuade the lawyer, Gandhi, to change his very public opposition to ten-print registration.

Later in the year, after Gandhi had broken off his negotiations with Smuts, he wrote to Asquith's government in Britain to explain the compromise. The Indians had agreed to give up their fingerprints 'only in order to enable the Government to have a scientific classification'.[48] In the history of satyagraha that he wrote and published in the 1920s Gandhi attributed the struggle to Lionel Curtis' clumsy enthusiasm for the 'scientific method'.[49] But in the months between January and June 1908, as he sought to encourage his own supporters to submit to fingerprint registration, it was Gandhi who became a passionate advocate of the scientific and progressive merits of fingerprinting.

In his writings in *Indian Opinion* in the month after the compromise Gandhi repeatedly berated his audience for their obsession with fingerprinting. But he simultaneously reversed his own course, presenting a fervent public case for full fingerprint registration as a technology of government. The most important of these statements is 'A Dialogue on the Compromise', which adopts, for the first time, the elastic Socratic style that he would use in *Hind Swaraj*.[50] If nothing else Gandhi's new-found

---

[45] Simon A. Cole, *Suspect Identities: A History of Fingerprinting and Criminal Identification* (Cambridge, MA: Harvard University Press, 2001), 174; the controversy over the scientific basis of latent identification continues unabated, Simon A. Cole, 'Is Fingerprint Identification Valid? Rhetorics of Reliability in Fingerprint Proponents' Discourse', *Law & Policy* 28, no. 1 (January 2006): 109–35. For Galton's demonstration of a one in four chance of a match between a single fingerprint and the entire global population, see Francis Galton, *Finger Prints* (London and New York: Macmillan and Co., 1892), 111.

[46] Cole, *Suspect Identities*, 174.

[47] 'Letter to Commissioner of Police', 19 June 1907, TAD LD 1466 Ag2497/07 Finger Impressions, 1907, TAD; Tennant, Secretary to Law Department to Assistant Colonial Secretary, Pretoria, 'Letter to Assistant Colonial Secretary', 19 June 1907, LD 1466 AG2497/07 Finger Impressions, 1907, TAD.

[48] British Indian Association, 'Petition to Secretary of State for Colonies', 9 September 1908, *Collected Works*, vol. 9, 119.

[49] Gandhi, *Satyagraha*, vol. 3, 127.

[50] Gandhi, 'A Dialogue on the Compromise'.

enthusiasm for the biometric state (elaborating on his earlier claims for the scientific virtues of thumb-printing) showed off his courtroom skills. Gone, now, was his outrage over the criminalising effects of ten-print registration, the emotional risks to the family, religious objections to the making of images, and the dangers to the honour of men. In their places, he presented an implausible case for the scientific merits of voluntary submission to ten-print fingerprinting. This argument – which his simultaneous and frantic written appeals for exemptions to Smuts showed as untenable and obnoxious – was the antithesis of the trenchant assault on scientific modernity that he began to articulate the following year (and which finds its completed form in *Hind Swaraj*).

In his attempts to persuade his audience of the virtues of fingerprinting, Gandhi echoed the arguments of Edward Henry and Francis Galton. 'The fact is that for the identification [of pass-holders] and for the prevention of fraud', he told his bewildered audience, 'digit-impressions offer a simple, effective and scientific means'.[51] The theme of the scientific, modernising character of fingerprinting provided the core of his new advocacy. He explained that 'finger-impressions are likely to be introduced everywhere sooner or later' because 'from a scientific point of view, they are the most effective means of identification'.[52] Fingerprints, he suggested, were 'a thousand times better' than the photographs which were being introduced on documents of identification in the Cape, because 'they cannot offend anyone's religious susceptibilities'.

To soften the points he had earlier made about the stigma of ten-print fingerprinting, which was used in India and in Britain only to identify criminals,[53] Gandhi pointed to the long experimental relationship between science and the prison system. When Jenner discovered the smallpox vaccine he first tested it on prisoners before offering it to the public. 'No one could argue', Gandhi claimed, 'that the free population was thereby humiliated'. Now that fingerprints had been freed from the 'enslaving law' that targeted Indians as a race, he urged his readers to embrace them because of their 'advantages from a scientific point of view'.[54] These recommendations, and the prospect of Smuts' draconian sanctions, worked well. During February 1908 the officials of the Asiatic Registry were overwhelmed by applications. At the end of the month

---

[51] M.K. Gandhi, 'Johannesburg Letter', *Indian Opinion*, 29 February 1908, *Collected Works*, vol. 8, 169.
[52] Gandhi, 'A Dialogue on the Compromise'.
[53] M.K. Gandhi, 'Letter to Colonial Secretary', *Indian Opinion*, 7 October 1907.
[54] Gandhi, 'Johannesburg Letter', 29 February 1908.

Gandhi reported that 'about 95 per cent of the Indians have already given their finger-impressions'.[55]

### Rejecting the soulless machine

In the months after his public agreement to begin voluntary registration Gandhi began to realise that Smuts would not repeal the Black Act (No. 2 of 1907) that specifically targeted Indians for fingerprint identification, nor did he intend to allow exemptions for a small number of educated Indians under the new immigration law. After frantically encouraging his supporters to undergo registration for the first half of 1908, Gandhi finally announced the resumption of resistance at a mass meeting held on 16 August. Thousands of people gathered at the Fordsburg Mosque to offer up the paper certificates of their voluntary registration to be symbolically burnt in a large cast-iron potjie. But the celebratory mood that greeted Gandhi's return to opposition to the Act did not last. Faced with a barrage of penalties and punishments and a system of identification that left them naked in the face of the Asiatic Registry's grasp, opposition collapsed utterly in the early months of 1909.[56] Gandhi's *Hind Swaraj* manifesto was famously the product of this defeat. It was written later in 1909, as Swan observed, while he was 'weighed down by the inadequacies of the first passive resistance movement'.

Even before the August mass meeting the state had begun to imprison people who did not produce registration certificates. Unlike the struggle a year earlier the officials were equipped with a complete set of fingerprint records of the Indians in the Transvaal which they could use to identify immigrants and resisters, their trading licences and property.[57] Early in the next year deportations began in earnest. The combination of fines, deportations, imprisonment, prohibitions on trading and property seizure left the Indians in the Transvaal with everything to lose and, having already offered up their fingerprints, nothing to gain. By May the 'vast majority of Transvaal Asians had emphatically repudiated the path of resistance'.[58] Gandhi's own political views also changed. As his constituents placed their economic assets before their own moral and political dignity he turned away from the merchant elites that had been the core of his concern before 1908 and, as Dhupelia-Mesthrie has observed,

[55] *Ibid.*     [56] Swan, *Gandhi*, 175, 173–8.

[57] Chamney, Registrar of Asiatics, Department of the Interior to Acting Secretary for Justice, 'Confidential', 23 April 1912, CAD JUS 0862, 1/138, 1910.

[58] Swan, *Gandhi*, 175, see also 173–8 for the details of defeat.

he 'began to see any prosperity amongst Indians at this time of struggle as a sign that they had become "accomplices ... in tyranny"'.[59]

By the end of 1909 popular resistance on the Witwatersrand, the source of so much hope and strength, had dwindled to a handful of die-hard *Satyagrahis*, many of them targeted for deportation. In the months immediately preceding the *Hind Swaraj*'s composition he had been locked in negotiations with the representatives of the Transvaal Boers in London. Smuts had, yet again, rejected his conservative proposals to make space for the principle of non-racial legal equality in the new South African constitution. And in Natal, open conflict had broken out between the (mostly Hindu) ex-indentured small-scale farmers and the (mostly Muslim) merchants.[60] Everywhere he turned his South African project was in ruins. One result was the retreat of the faithful to Tolstoy Farm, outside of Johannesburg. A growing emphasis on the moral, and interior, qualities of satyagraha, valuing personal integrity above material benefit and political advantage, was another. But Gandhi also began to deploy the apocalyptic language of the social revolutionary, rejecting in sweeping and contemptuous terms the foundations of the English liberal civilisation that he had defended so vigorously before 1908. The origins of this militant rejection of Western modernity lay in his bitter experience of the compromises over fingerprint registration.

It was Carlyle's state as machine,[61] and the horrifying effects of a normative physics that sought to treat human beings as machines, that possessed Gandhi from the middle of 1908. 'Machinery is the chief symbol of modern civilisation', he says in *Hind Swaraj*, 'it represents a great sin'. Like the Luddites he worried about the destructiveness of modern technologies, but when he claimed that machinery 'has impoverished India' he had in mind a much wider cultural and religious decline.[62] This obsession was unmistakably taken from his reading of Ruskin's work *Unto this Last*, which he began to expound upon in detail in the last weeks of May, as the compromise with Smuts was disintegrating. For Ruskin the central problem of modern English political life was the application of a Smithian physics to the government of human beings. 'Let us eliminate the inconstants', he mocked the political economists, 'and, considering the human being merely as a covetous machine, examine by what laws

---

[59] Uma Dhupelia-Mesthrie, *Gandhi's Prisoner: The Life of Gandhi's Son Manilala* (Cape Town: Kwela Books, 2007), 77.

[60] *Ibid.*, 177, 200; Lake and Reynolds, *Drawing the Global Colour Line*, 217; Bhana and Vahed, *Gandhi in South Africa*, 67.

[61] Jon Agar, *The Government Machine: A Revolutionary History of the Computer* (Cambridge, MA: MIT Press, 2003), 37–8.

[62] Gandhi, *Hind Swaraj*, 107.

of labour, purchase, and sale, the greatest accumulative result in wealth is obtainable'.[63] Raging against the obvious moral effects of this capitalist market as a guide to the good, Ruskin insisted that the human being was 'an engine whose motive power is a Soul'. This claim, that Western modernity had abandoned its metaphysical compass in its frantic search for material prosperity, became the key argument of Gandhi's political philosophy.

In May 1908 he chose the title *Sarvodaya*, or the advancement of all, for his detailed Gujarati paraphrasing of *Unto this Last* in *Indian Opinion*. And his introduction frames Ruskin as the most important English critic of Benthamite utilitarianism, dismissing the claim that the goal of society should be the 'the happiness of the greatest number', even 'if it is secured at the cost of the minority' and in violation of divine law.[64] Two months later, when he had completed the nine-part translation and the compromise with Smuts had collapsed into bitter conflict, his explanation of the significance of Ruskin's work had expanded into a wholesale rejection of modern capitalism.

All of the arguments of *Hind Swaraj* are presented in summary in this concluding comment, with a much more direct link to the lessons of the political struggle in South Africa than anything in the book. Writing a year before his encounters with the young radicals in London in 1909,[65] he warned the enthusiasts of violence that 'the bombs with which the British will have been killed will fall on India after the British leave'. He reminded his readers that despite the comparative youth of European civilisation it had already been reduced 'to a state of cultural anarchy' and stood on the brink of a terrible war. And he asked them if the sovereignty they hankered for was the kind that Smuts had secured for the Transvaal. Smuts who 'does not keep any promise, oral or written' represented a political elite 'who serve only their own interests' and who 'will be ready to rob their own people after they have done with robbing others'. Real sovereignty for India would follow from the number of citizens who chose to live a moral life, and it 'will not be possible for us to achieve it by establishing big factories' or through the 'accumulation of gold and silver'. All this he said, very generously, 'has been convincingly proved by Ruskin'.[66]

---

[63] John Ruskin, *Unto This Last* (London: Cornhill Magazine, 1860).
[64] M.K. Gandhi, 'Sarvodaya', *Indian Opinion*, 16 May 1908, *Collected Works*, vol. 9.
[65] See Parel's insightful account of the origins of Hind Swaraj in Gandhi, *Hind Swaraj*, xxxvii–xli.
[66] M.K. Gandhi, 'Sarvodaya [-IX]', *Indian Opinion*, 18 July 1908, *Collected Works*, vol. 8, 455.

A key part of this critique, much of which would not have been imaginable by Ruskin writing in the 1850s, was a fierce rejection of the benefits of the technologies of the second industrial revolution. When Gandhi was called to debate the question 'Are Asiatics and Coloured races a menace to the Empire' at the Johannesburg YMCA on 18 May 1908, as Smuts' refusal to withdraw the Black Act was becoming obvious, he spent much of his time rebutting the claims of the new 'segregation policy'. Speaking in the city of gold, surrounded by the workshops, railways and reduction works of the largest mines in the world, Gandhi warned his audience that 'South Africa would be a howling wilderness without the Africans'. Segregation, he suggested, was a product of the ascendancy of the Spencerian moral philosophy of the 'survival of the fittest', which made physical and intellectual strength the means and the end of Western civilisation. 'I decline to believe', he told the audience at the YMCA in terms that were elaborated in *Hind Swaraj*, 'that it is a symbol of Christian progress that we have covered a large part of the globe with the telegraph system, that we have got telephones and ocean greyhounds, and that we have trains running at a velocity of 50 or even 60 miles per hour'. Machines, on the contrary, had become the measure, and the telos, of a brutal, un-Christian, goal-less, progress, and the essence of 'western civilisation'.

By the 1930s Gandhi had come to the view that modern bureaucracy was an instrument of genocidal conflict. The state, he argued, was a 'soulless machine' which 'can never be weaned from violence to which it owes its very existence'.[67] The real danger, for India, was that progressive government threatened to introduce the state into every village and every home. He chastised the economic historian, Manmohan P. Gandhi, for suggesting that there were similarities between the character of violence under the Mogul state and its British colonial heir. 'Formerly, the [Mogul] government touched the lives of only those who were connected with the administrative machinery', Gandhi responded, anticipating Foucault's argument about the new forms of governmentality: 'It is only in the present age that governments have become eager to extend their grip over entire populations.' In this project of universal administration it was British rule that had 'acquired the utmost efficiency' and which posed the most serious danger to India.[68]

---

[67] M.K. Gandhi, interview by Nirmal Kumar Bose, 9 November 1934, *Collected Works*, vol. 65, 316.
[68] M.K. Gandhi to Manmohandas Gandhi, 'Letter to Manmohandas P. Gandhi', 3 March 1930, *Collected Works*, vol. 48, 371.

Many scholars have noted the profound shift in Gandhi's politics in South Africa during 1908.[69] In his own accounts the turn was a shift away from the external world to the interior self, from politically to personally motivated change, but there is an element of deception here. In South Africa in 1913, and afterwards in India, Gandhi's interest in personal transformation was always attached (although the connection was often frayed) to the wider political force of mass protest. If anything, after 1908, Gandhi was more attentive to the ideological demands of maintaining a mass constituency than he had been before. His self-conscious adoption of the tactics of saintliness, deploying the life-threatening fast as a weapon against his friends and his enemies, ensured that he was almost invulnerable to the charges of corruption and self-interest his critics had used against him in 1908.[70] The change was certainly not a dramatic disavowal of the British empire. Gandhi's views of the moral and political virtues, and failings, of the British empire remained strikingly and consistently ambiguous from the 1890s into the 1930s. Nor was it a fundamental realisation and rejection of the simplistic forms of racism coursing through the empire in this period. Gandhi's broadly Victorian and paternalistic views about the civilisational prospects of different races remained with him after 1908.[71]

The real change was in his understanding of the nature and purpose of the state. Before 1908, he had seen the state as an instrument of harmony, shaped by science and law, and he had understood his own practice as an extension of that power. Afterwards he viewed the 'administrative machinery', with its technological means and telos, as an instrument of destruction.

It was Gandhi's unique understanding of the vulnerability of the state's power that underpinned his self-righteous explanation of his own participation in the making of the fingerprinting regime in the Transvaal. His insistence that the state was constrained by the cooperation of its subjects, no matter how refined in law or administration, equipped him with a serene confidence in the limits of bureaucratic power, and the normative superiority of non-violence, that he maintained to the end of his life. As the events of the century took on ever more appalling forms of state violence he maintained his quixotic, and increasingly metaphysical,

---

[69] Bhana and Vahed, *Gandhi in South Africa*; Prozesky and Brown, *Gandhi and South Africa*; Gandhi, *Hind Swaraj*; Brown, *Gandhi*; Swan, *Gandhi*.
[70] Brown, *Gandhi*, 311.
[71] 'Adminstrative inequality must always exist so long as people who are not the same grade live under the same flag.' M.K. Gandhi, 'Letter to "The Star"', 17 September 1908.

conviction in the potentials of non-cooperation, famously recommend-
ing to the Czechs and the Jews in Europe in the 1940s that they should
follow the example of the Indians in South Africa in adopting non-violent
resistance.[72] This confidence in the ultimate moral and political superiority of non-violence accounts for the absence of anything resembling
an acknowledgement of his own tactical mistakes, yet the timing and the
vocabulary of his writing from May 1908 suggest that they were deeply
felt. It was the effect of his horrible political disappointment with the
results of the fingerprinting compromise that prompted him to frame an
abruptly antithetical understanding of the politics of law and science in
government.

The politics of satyagraha, in South Africa and afterwards, rested upon
a massive simplification: 'where there is or there is not any law in force,
the Government cannot exercise control over us without our cooperation'.[73] The key principle of the supremacy of withholding cooperation,
which he was to put to good use in Natal in 1912 and again in India as
the vehicle of Congress's nationalist legitimacy,[74] did not work under
the special conditions of biometric registration in the Transvaal. Once
offered, the full fingerprint registrations provided in February 1908
could never be withdrawn.

Under the fingerprint requirements Indian immigration in to the
Transvaal dried to a trickle after 1908. The state carefully maintained a
complete set of biographical files on every male Indian in the Transvaal
from this period up to the late 1970s which regulated the property holdings, business activity, rights of domicile and family memberships. The
original fingerprints offered under the terms of the compromise provided an anchor of control that set an ominous precedent for the form
and ambitions of the South African state.

## Conclusions

The global significance of *Hind Swaraj*, as the ideological framework of
the political strategies of satyagraha in India, can scarcely be overstated.
It was, as Chatterjee has shown, a spectacularly successful movement of
paradoxes: 'a nationalism which stood upon a critique of the very idea
of civil society, a movement supported by the bourgeoisie which rejected
the idea of progress, the ideology of a political organization fighting for
the creation of a modern national state which accepted at the same time

[72] Brown, *Gandhi*, 321.
[73] Gandhi, *Satyagraha*, vol. 3, 217 18.
[74] Brown, *Gandhi*, 109–14, 219–20 and (the famous salt-march), 238–9.

the ideal of an "enlightened anarchy".[75] And, while Gandhi's rejection of the state had little practical influence on the Nehruvian state, or on the contemporary plans for biometric government in India, the ideological power of his arguments against imperial progressivism in the anticolonial protests of the twentieth century, and even up to the present, has been profound.

And while it is also true that the fingerprint registrations that Gandhi advocated in early 1907 set up a comprehensive register of identity for Indians in the Transvaal, his protests also taught Smuts and many others of the dangers of a globally coordinated campaign of protest and resistance. As we will see in the following chapter, this experience placed meaningful limits on the plans and ambitions of the advocates of biometric government.

---

[75] P. Chatterjee, *Nationalist Thought and the Colonial World: A Derivative Discourse* (London: Zed Books, 1986), 101, on the general significance of the book, 84–125.

# 4    No will to know: biometric registration and the limited curiosity of the gatekeeper state

This chapter follows the development and dismantling of civil registration for rural black people in South Africa in the decades after Gandhi left South Africa. This sounds terribly dull, but the story is intriguing and important in several ways. The South African state was not poor, and, unlike its northern neighbours, it was not bereft of administrative capacity. Perhaps the best indicator of this wealth and bureaucratic strength was the extension after 1944 of means-tested old-age pensions to black South Africans.[1] By 1950 the state was spending over a million pounds a year on 200,000 African pensioners.[2] Dozens of different registration systems were also developed by the state in this period. To name only the largest of these: by the 1930s the Native Affairs Department (NAD) had in place a centralised system of tax registration that, in theory, recorded a name, identity number and address for every African man over the age of eighteen in the country.[3] More provocative is the fact that the system of civil registration was actually working well in the Eastern Cape and in Natal in the early years of the twentieth century. It was deliberately foreclosed and then abruptly abandoned in 1922. But perhaps the most intriguing problem derives from that fact that historians of the South African state, and society, have characterised it as knowledge driven.

An earlier version of this chapter was published as 'No Will to Know: The Rise and Fall of African Civil Registration in 20th Century South Africa', in *Registration and Recognition: Documenting the Person in World History*, ed. Keith Breckenridge and Simon Szreter, Proceedings of the British Academy 182 (Oxford University Press, 2012), 357–85.

[1] Jeremy Seekings, 'Visions, Hopes & Views about the Future: The Radical Moment of South African Welfare Reform', in *South Africa's 1940s: Worlds of Possibilities*, ed. Saul Dubow and Alan Jeeves (Cape Town: Juta, 2005).

[2] Alan Rycroft, 'Social Pensions and Poor Relief: An Exercise in Social Control' (Unpublished, 1987); Andreas Sagner, 'Ageing and Social Policy in South Africa: Historical Perspectives with Particular Reference to the Eastern Cape', *Journal of Southern African Studies* 26, no. 3 (September 2000): 540.

[3] F.R. for Secretary for Native Affairs to Director of Census and Statistics, Pretoria, 'National Register of Population', 12 October 1949, SAB NTS 9360 2/382 Part 2 Registration of Births and Deaths (natives) in the Union. (general File). 1947 to 1952.

Why, then, did it choose not to gather the most basic kind of administrative information?

In Egypt and in India the English colonial state in the nineteenth century battened on to already existing written systems of land and identity registration, for revenues and recruitment, while simultaneously complaining bitterly of the inadequacies and corruption of indigenous systems of government.[4] Africans lived in a world in which these forms of identity registration would have been absurd, one in which land was plentiful and people were always alarmingly scarce; this was a political and emotional world in which every individual was carefully and concretely located within multiple, overlapping and always contradictory forms of kinship affiliation – age-regiments, houses, descendants, affines, mothers' relatives, fathers' relatives. All the primary forms of property in this world – which generally did not include land itself – were also carefully attached to the management of these kinship relationships.[5] In this world there was no administrative bounty available like the Egyptian and Indian systems, and the colonial state sought, famously, to mobilise other forms of local authority in the defence of its power.

By the start of the twentieth century the blunt and blind administrative instrument of communal land tenure controlled by chiefs or headmen had become for colonial officials, and for many Africans, the preferred mechanism for a welfare net.[6] The conventional explanation that historians have offered for the prevalence of this system of 'hegemony on a shoe-string' has been the systematic parsimony of the colonial state.[7] A reluctance to spend money on Africans in the countryside certainly contributed to the decline of civil registration but there were other, more important, intellectual motivations. In this chapter I want to show that the administrative inadequacy of indirect rule in South Africa followed from the contested and constrained character of progressive ideas about the state in the decades after 1910. In short, a profound disagreement between those who sought to use scientific forms of policing – and

---

[4] Alfred Milner, *England in Egypt* (London: E. Arnold, 1892).

[5] Jan Vansina, *Paths in the Rainforest: Toward a History of Political Tradition in Equatorial Africa* (Madison: University of Wisconsin Press, 1990), 146–55; Jean Comaroff and John L. Comaroff, *Of Revelation and Revolution: The Dialectics of Modernity on a South African Frontier*, vol. 1 (University of Chicago Press, 1991), 150–69; S.O. Kwankye, 'The Vital Registration System in Ghana: A Relegated Option for Demographic Data Collection', in *The African Population in the 21st Century* (presented at the Third African Population Conference, Durban: Popline, 1999), discusses this in contemporary Ghana.

[6] Lungisile Ntsebeza, *Democracy Compromised: Chiefs and the Politics of Land in South Africa* (Cape Town: HSRC Press, 2006), 74–5; Sagner, 'Ageing and Social Policy in South Africa', 560.

[7] Sara Berry, 'Hegemony on a Shoestring: Indirect Rule and Access to Agricultural Land', *Africa: Journal of the International African Institute* 62, no. 3 (1992): 327–55.

particularly fingerprinting directed at adults – confronted the advocates of scientific public health informed by voluntary reporting. The primary result of this conflict was administrative inertia.

The failure of African civil registration in the inter-war period in South Africa followed from a particular political arrangement. Administrative concern for the registration of births and deaths was, as Foucault might suggest, repeatedly demanded by the advocates of a system of public health, from 1900 through to the end of the 1940s. These progressive reformers were opposed, especially in Natal, by full-blooded opponents of bureaucratic intervention – key amongst them the wealthiest owners of the sugar plantations who also controlled the government after 1893. Later civil registration was also opposed by some of the state-appointed chiefs, and by the most pessimistic liberal magistrates defending personalised government.

But the fatal weakness of the system of birth registration in South Africa stemmed from the fact that it faced a competing scheme for identity registration. From the early years of the twentieth century another group of progressive administrators – advocates of control from the mining industry, the police and the Pass Offices of the NAD – proposed compulsory universal fingerprinting of all African men as an alternative to universal civil registration. This obsession with fingerprinting as an instrument of central government control, as we have seen, began in earnest when Edward Henry set up the Transvaal Police in 1900. In the generation after Smuts' struggle with Gandhi the advocates of universal biometric registration faced determined liberal scepticism in the state, and it was only in the 1950s, under Apartheid (as I will explain in the next chapter), that the state finally completed the early twentieth-century plans for the compulsory fingerprint registration of all African adults.

In the inter-war period the advocates of biometric registration succeeded only in maligning the value of the information that was produced by delegated civil registration, and that provided the rationale for the system's dissolution. It was this unwieldy conflict between the policing and the public health branches of progressive government, combined with the absence of meaningful political representation, that, in the end, meant that the state consciously chose not to gather the most basic information about its African subjects.

## A progressive deviation

The gathering of birth and death information from black people in South Africa began falteringly in Natal in the last years of the nineteenth century. Much earlier, in 1858, the colonial legislative council had proposed a census and registration of the African population, as part of the effort to coerce labour from Shepstone's tribal reserves. That proposal was rejected by the Lieutenant Governor as 'unnecessary interference'.[8] A new registration law was passed in 1896, aimed at the *Amakholwa* – the thousands of Africans married according to Christian rites and thus lying between the 5,000 in Natal who were legally exempted from the operations of customary law and the 350,000 who lived under chiefly government. Possession of an official Letter of Exemption essentially conferred on Africans some of the key rights of British citizenship. It entitled women to register (and bequeath) freehold property, and to access British courts; for men it allowed the vote. But the law demanded that an application for exemption could only be made by Africans who had passed through all of the requirements of mission life (monogamy, literacy, good character) and it was always fiercely policed in Natal. Importantly, children inherited their parents' status only if they were born prior to the issuing of the Letter of Exemption.[9] It was for this reason that the 1896 Act required Christian parents to register the birth, and death, of their children within thirty days, or face a £5 penalty. The Act specifically placed the burden of the registration – of both birth and death – on the parent, which suggests that it was concerned solely to limit the population that might claim exemption from customary law.[10] This measure also allowed the members of the new Responsible Government, who had previously been critics of the imperial government's adoption of Theophilus Shepstone's system of benign neglect, to do something, without actually stirring up change or, critically, incurring any expense.

Between 1846 and 1877 Shepstone had been almost solely responsible for the government of the African population of Natal, developing an ad-hoc paternalism that had powerful effects on the form of the African colonial state. During this period he moulded a distinctive form of administration which relied on appointed chiefs as the agents of an undocumented system of government, and a crudely effective hut tax

---

[8] In 1869 the colony required that every African marriage had to be registered (and recorded centrally) at a cost to the applicants of £5. There is little evidence that the scheme ever worked in practice. Benjamin Kline, *Genesis of Apartheid: British African Policy in the Colony of Natal, 1845–1893* (Lanham: University Press of America, 1988), 58; Norman Etherington, 'The "Shepstone System" in the Colony of Natal and beyond the Borders', in *Natal and Zululand from Earliest Times to 1910: A New History* (Pietermaritzburg: University of Natal Press and Shuter & Shooter, 1989), 174; John Lambert, *Betrayed Trust: Africans and the State in Colonial Natal* (Pietermaritzburg: University of Natal Press, 1995), 46.

[9] Edgar Brookes, *The History of Native Policy in South Africa from 1830 to the Present Day* (Cape Town: Nasionale Pers, 1924), 197–8.

[10] Secretary for Native Affairs, *Circular: Registration of Birth*, vol. 3/1/2, 1/KRK, 1898.

as the main source of colonial revenues. In return Shepstone worked hard, and effectively, to resist the settlers' claims to African land and labour. Under his system, the chiefs were supervised by a handful of appointed magistrates, each responsible for very large districts, several chiefs and thousands of homesteads. The title magistrate may be misleading. In the nineteenth century the individuals who were appointed as Native magistrates rarely had anything resembling legal training. In the following century the posts were typically filled, in Natal as in the rest of South Africa, by individuals who had distinguished themselves as officials in the NAD – in both centuries a mission background was much more important than legal experience.[11] The magistrate's work consisted, very largely, of gentle supervision of the chiefs in whose hands the daily routine of government rested unmolested. Shepstone's system of 'native agency' became synonymous, especially after 1900, with resisting cultural and economic change. (It was Welsh who first argued that it was Shepstone's method of indirect rule – which he extended to the Transvaal after 1877 – that provided the administrative and intellectual seedbed of Apartheid, and, more recently, Mamdani who suggested that it provided the basis of colonial rule across the continent.)[12]

By the turn of the century almost all of the local settler leaders had become enthusiasts for the doctrine (now habitually attributed to Shepstone) that the state should interfere as little as possible with African society.[13] The prime minister, Sir Hercules Robinson, a fierce critic of post-Shepstonian 'drift' before 1893, discovered the special virtues of ignorance after he took responsibility for government. The Colony, he explained in the legislature, 'is to be congratulated that there is so little occasion to give information regarding native affairs, because it shows that the Native population is generally contented and that they give the Government very little reason for consideration or reflection'.[14]

Natal's enthusiasm for administrative drift lapsed momentarily during Milner's progressive remaking of the Boer Republics. In a flurry of legislative energy the colony passed a string of new capitation laws aimed

---

[11] Saul Dubow, *Racial Segregation and the Origins of Apartheid in South Africa, 1919–36* (Oxford: Macmillan in association with St Antony's College, 1989), 99–101.

[12] Etherington, 'The "Shepstone System" in the Colony of Natal and beyond the Borders'; David John Welsh, *The Roots of Segregation: Native Policy in Natal (1845–1910)* (Cape Town: Oxford University Press, 1971); Peter Delius, *The Land Belongs to Us: The Pedi Polity, the Boers and the British in the Nineteenth Century Transvaal* (Johannesburg: Ravan Press, 1983); Mahmood Mamdani, *Citizen and Subject: Contemporary Africa and the Legacy of Late Colonialism* (Cape Town: James Currey, 1996).

[13] Welsh, *Roots of Segregation*, 229–31; Brookes, *Native Policy in South Africa*, in contrast, sees the Shepstone system as a model of energetic segregationism.

[14] Welsh, *Roots of Segregation*, 229.

at tightening the colonial state's grip on its individual African and Indian subjects. First amongst these was a new Pass Law, Act 49 of 1901 for the 'Identification of Native Servants', which adopted the passport, name registration and labour distribution elements from the Pass Law drafted by Jan Smuts in the Transvaal in 1899, but without the revenues that were raised there from employers. Following in the same vein as the earlier law that required converts to register births, in 1903 the Colony extended the requirement that contract-expired indentured Indian migrants pay an annual £3 tax to include their teenage children. Two years later, as the colony struggled to replace the custom revenues lost in the recession that followed the withdrawal of British troops, the Treasury pushed for a change in the organisation of the taxes levied on Africans, moving from Shepstone's carefully designed hut assessment to a poll tax levied on all adult men in the colony, black and white. Although the new law exempted homestead heads who were already paying the hut tax it struck at the heart of the old homestead political economy and pushed some of the Natal chiefs into rebellion.[15]

Hidden beneath these taxes was a capitation law of a different kind. In 1901, Natal passed a Public Health Act that required, amongst other progressive measures, the appointment of a colonial Health Officer and the establishment of the infrastructure of a basic health service. As Marcia Wright has shown, this fledgling health system was designed by a member of the Indian Medical Service who was brought to Natal in 1899 to act as Special Plague Adviser, bringing with him the experience of the Indian Civil Service's (ICS) effort to impose compulsory birth registration in the 1890s.[16] From the outset this service faced fierce and determined opposition from the leading figures of the settler government. Chief amongst these was Sir Liege Hulett, one of the most important planters, who as Wright observes, 'growled that the Health Department was too expensive, not worth its cost'. It was Hulett – an outstanding Methodist,

---

[15] Shula Marks, *Reluctant Rebellion: The 1906–8 Disturbances in Natal* (Oxford: Clarendon Press, 1970), 131–43; Jeff Guy, *The Maphumulo Uprising: War, Law and Ritual in the Zulu Rebellion* (Pietermaritzburg: University of KwaZulu-Natal Press, 2005), 23–4; Lambert, *Betrayed Trust*, 167; Benedict Carton, *Blood from Your Children: The Colonial Origins of Generational Conflict in South Africa* (Pietermaritzburg: University of Natal Press, 2000), 118; Andrew Duminy, 'Towards Union, 1900–10', in *Natal and Zululand from Earliest Times to 1910: A New History*, ed. Andrew Duminy and Bill Guest (Pietermaritzburg: University of Natal Press, 1989), 402–23.

[16] Marcia Wright, 'Public Health among the Lineaments of the Colonial State in Natal, 1901–1910', *Journal of Natal and Zulu History* 24 (2007); the Durban municipality had, for decades, tried, without much success, to use sanitation laws against Indian landowners: Maynard W. Swanson, '"The Asiatic Menace": Creating Segregation in Durban, 1870–1900', *The International Journal of African Historical Studies* 16, no. 3 (1983): 401–21, called this the 'sanitation syndome'.

liberal patron of John Dube, and founder of Kearsney College – who led the ongoing effort to limit the costs and retrench the services that were planned in 1901.[17]

The first goal of the newly appointed Health Officer was to extend the comprehensive system of birth and death registration that was maintained for the indentured migrants to the wider population of Africans and Indians. 'The basis of all work for the improvement of the Public Health is Vital Statistics', he explained, and 'without strictness of registration no reliable statistics can be obtained'.[18] This effort to establish a unitary system of civil registration in Natal failed in the face of determined opposition from the planters and from local doctors, but the fleeting enthusiasm for a public health system had one significant legislative effect: Act 25 of 1902 required the registration of all African births and deaths in the old colony of Natal, excluding the territory of Zululand.

The scheme for civil registration in Natal had some typically Shepstonian features. Under the law, responsibility for the notification of births and deaths fell on to the male 'kraal head'. He was required, within three days, to report the event to the district Official Witness. These unsalaried positions had been created under the Natal Native Code of 1891 to sanction customary marriages, and they were rewarded through an increased allowance for *ukulobola*. The Official Witness was required to report all such events to the local magistrate within thirty days. For every recorded event – births and deaths – Official Witnesses were paid one shilling by the magistrate. This was a substantial fee at the time, amounting to the better part of a day's wage, and clearly the reason for the scheme's relatively rapid introduction. The legislation specifically prohibited the registration of stillbirths, but required that 'in the case of children dying shortly after birth, both the birth and death must be registered'. Parents were permitted to report the names of children who died unnamed at a subsequent date 'for insertion in to the register and the completion of the record'.[19] The new system did not cover all Africans. Couples married under Christian marriage rites remained outside of this system, and along with the Exempted, their births and deaths were incorporated in the register maintained by the colonial (and later provincial) registrar.

[17] Wright, 'Public Health among the Lineaments of the Colonial State in Natal', 161.

[18] *Ibid.*, 159.

[19] Magistrate, Kranskop, '1/KRK 3/1/4, KK931A/1902 Government Notice No. 799, 1902: Rules Re Registration of Births and Deaths of Natives', 1902, NAB; on the role of the official witnesses, see Thomas V. McClendon, 'Tradition and Domestic Struggle in the Courtroom: Customary Law and the Control of Women in Segregation-Era Natal', *The International Journal of African Historical Studies* 28, no. 3 (1995): 527–61.

The convulsive expansion of administrative interventions, taxes and registration, combined with the collapse of the war boom, the enduring effects of the Rinderpest epidemic of 1897, and a dramatic increase in population made life very difficult for Africans in Natal. Discontent culminated in the last days of the summer of 1906 in widespread protests and refusal to pay the new poll tax, and isolated acts of violence and ritual killing. After the rebellion had been brutally suppressed by colonial irregulars who took the opportunity to show off the skills in scorched-earth warfare they had learned against the Boers, the colony launched a grand commission, in July 1907, to consider the grievances of its African subjects.[20] The committee, which included Native Experts like Hulett, Maurice Evans and James Stuart, heard evidence from over 5,000 African witnesses; the report of these sittings is eloquent testimony to the discontent and distress of the household patriarchs, and their chiefs, in the face of the unworkable combination of the newly rationalising state and Shepstone's reliance on 'Native agency'.

The senior men who came to address the commission protested bitterly about the new instruments of tax and surveillance; they worried about the moral effects of young men paying their own tax (and retaining their own tax receipts), complained about the increase in the hut tax to fourteen shillings, of the dog tax and the stringent methods being used to control the movement and dipping of African-owned cattle. They fumed about the ways in which the old Shepstonian order was being deformed by white officials' expectations that they all be treated as chiefs, of the collapse of the careful practice of consultations and the ongoing decline in the prestige of the chiefs. And they protested at a host of new dangers: lawyers' fees, sex between whites and blacks, inadequate wages, the fencing off of private lands, and usurious money lending.

Amidst this litany of protest the vital registration requirement was an insignificant source of grievance, but four men, amongst the hundreds who gave testimony and the thousands who attended the hearings, grumbled about the new law. The first to introduce the issue was Nduku from the Klip River Division in northern Natal. He complained that he had recently 'experienced much trouble in regard to the registration of births and death' and that he 'was punished if he did not register his child within the proper time'. He particularly objected to the requirement to provide a name hastily.[21] Mzungulu, from Krantzkop, also complained of the requirement to name the child 'forthwith', and he asked, 'Why

---

[20] Guy, *The Maphumulo Uprising*; Brookes, *Native Policy in South Africa*, 77.
[21] Natal. Native Affairs Commission, *Evidence [of The] Native Affairs Commission, 1906–7* (Pietermaritzburg: P. Davis, 1907), 736.

should a man for not reporting a death, be suspected of having committed a criminal offence?'[22] Faku, an *iNkhosi* from Ixopo, observed amidst a list of wider complaints that he was 'surprised' by the registration requirement.[23] This little stream of evidence was hardly damning, but one witness, Chief Sibewu of the Lower Umzimkulu Division, near Port Shepstone, made a much stronger point. His only comment was on civil registration, which he said was unnecessary, and he warned the committee that the people worried 'what the intention of the Government was when the children grow up' and that 'they felt it would get hold of them, and send them off to some other place'.[24]

The report that was produced from the 1907 Commission generally addressed the spirit and not the letter of the new laws regulating rural African families. It argued, for example, for more respectful treatment by the magistrates and their police of homestead heads, and for a more paternalistic approach to the attestation of the promissory notes that were the source of entangling debts. But, providing a classic example of the ways in which the segregationists sought out the idiom of African reaction as the justification of their own plans, the only specific legislative recommendation made by the committee was for the repeal of Act 25 of 1902 'having regard to Native feeling thereon, the incompleteness of records, the exclusion of Zululand from its operation, and the cost of such registration'.[25] The legal requirement for civil registration was not withdrawn in 1907, but the commission's recommendation that it be abandoned hung like a Damoclean sword over the scheme for the remainder of its life.

## Advance or retreat

Immediately after the formation of the Union in 1910 the new central government in Pretoria began to consider the question of how to move forward on the project of civil registration for Africans.[26] In the Cape, compulsory civil registration had been introduced in the Transkeian Territories in 1899, but without the two critical elements that made the Natal scheme work: payment for local registrars and punishment

---

[22] *Ibid.*, 856.    [23] *Ibid.*, 782.    [24] *Ibid.*, 808.
[25] Brookes, *Native Policy in South Africa*, 85.
[26] Colonial Secretary, 'NAB CSO 1893, 1910/4799 Secretary for Interior Pretoria Asks for Memorandum Shewing System Prevailing in Natal for Registration of Births and Deaths, Particularly Natives', 1910, Pietermaritzburg; Colonial Secretary, 'NAB CSO 1902, 1911/1482 Secretary for the Interior Pretoria: Asks to Be Furnished with a Memorandum Under Which the Systems of Registration of Births, Marriages and Deaths Is Administered', 1911.

for delinquents. In 1911 the national Department of the Interior (DOI) complained about the desultory character of the data from the Transkei. The annual number of registrations, which at no time represented anything like the actual number of births and deaths, had declined steadily, reaching a nadir of 4,000 in 1910. Smuts' officials in Pretoria worried – with their hands already busy with the registration struggle with Gandhi – that the attitude of parents in the Transkei was 'either utterly apathetic or one of passive resistance'. And they wanted the issue laid before the Bunga so that 'the free discussion in all its bearings might result in the enlistment of the active cooperation of the Native Members of the Council'.[27] There was, at this stage, no question of abandoning the scheme. Instead, Stanford, the Chief Magistrate, announced the start of prosecutions for failing to register births and deaths.[28]

The problem in the Cape was that registration, and prosecutions, were both the responsibility of poorly paid Headmen, or *Izibonda*.[29] 'The person to whom Government looks to see that the law is observed', Brownlee commented after prosecutions had begun, 'is the same person who is criminally liable for its observance and if he fails in his duty it can only be by accident that his omission in individual cases is discovered'. After the complaints from the DOI in 1911, the Transkei magistrates began to look for alternative ways of handling registration, and, after some debate, resolved that the most promising candidates for the position of local registrar were the teachers in government-funded schools. They then turned to Smuts' old department for the funds that would be required to reward teachers for taking on the processing of birth and registration forms – the suggested rate of payment had now fallen dramatically to between three and sixpence per event. In what must count as one of the world's finest examples of the old adage 'penny wise and pound foolish', the DOI refused to accept responsibility for the £800 budget that would be required to make civil registration work in the Transkei, arguing that the matter was being held in 'abeyance until a

---

[27] Acting Under-Secretary for the Interior to Dower, Under Secretary for Native Affairs, March 3, 1911, SAB NTS 9363 3/382 Transkeian territories: Registration of births and deaths, 1911–1960.

[28] S.H. Stanford, Chief Magistrate, Transkeian Territories to Dower, Secretary for Native Affairs, March 3, 1911, SAB NTS 9363 3/382 Transkeian territories: Registration of births and deaths, 1911–1960.

[29] For the special place of the headman (and the magistrates, and the Bunga) in the Transkei, see William Beinart and Colin Bundy, 'A Voice in the Big House: The Career of Headman Enoch Mamba', in *Hidden Struggles in Rural South Africa: Politics and Popular Movements in the Transkei and Eastern Cape, 1890–1930* (Cape Town: James Currey, 1987).

uniform law is passed when all parts of the Union can be dealt with on the same basis'.[30]

Accustomed as they were to administering a state without access to the funds they raised in taxes, the Cape magistrates changed their strategy, abandoning the idea that registrars should be paid for their work, and arguing instead for a change in the law in order to teach the moral lessons of registration. They argued, after 1913, that the regulations should be changed to place the burden of registration on to 'the owner [of the household where the vital event occurred] or person in charge for the time being'. This formulation, unlike the Natal system and the later requirements of civil registration under Apartheid, left open the possibility that women would be legally responsible for the registration of their own children or grandchildren. Under this arrangement, headmen would be confined to their familiar role as policemen, ensuring that heads of households did their duty.

The idea was submitted to the provincial registrar for comment, and he produced an assessment of civil registration in the Transkei that was, in the long run, to have powerful effects. He mocked the magistrates' request for relief for the overburdened and underpaid headmen, as 'with very few exceptions, the Headmen have not allowed the duty to weigh heavily upon them'. Working through the most conservative and undeveloped territories of Pondoland he pointed out that huge districts, like Bizana and Flagstaff, had not managed to register a single birth or death in the course of the previous year. Where registration was taking place in a more regular fashion it was typically 'made by relatives and friends and not by the Headmen'. Yet he held out very little hope that anyone else would be able to do the job any better. The headman, he pointed out, 'is constantly in touch with the Magistrate and has periodically to present himself at the Magistrate's office to draw his pay' where the ordinary head of household had hardly any contact with white officials. The registrar was also utterly sceptical about the magistrates' optimistic view that unpaid storekeepers would 'assist the natives, their customers, by completing information forms for them'. But he was mildly enthusiastic about the broader project of presenting a moral lesson. As the new arrangement, placing legal responsibility on the homestead head, 'would at least serve the useful purpose of introducing to the Natives the idea of individual responsibility for reporting births and deaths, which it is intended to enforce later, I think the experiment should be tried'. The Cape law was duly altered

---

[30] Secretary for Interior to Secretary for Native Affairs, Pretoria, 15 October 1913, SAB NTS 9363 3/382 Transkeian territories: Registration of births and deaths, 1911–1960.

in September 1914, but no additional funding was allocated to registration, and the dismissive assessment of the quality of the data was what remained in the record.

A similar problem shaped the development of the system in Natal. Here the problem was not the infrequency or irregularity of returns (a problem that was manifestly solved by the one shilling payment to Official Witnesses), it was the fact that the entire territory north of the Thukela river was not covered by the scheme. After 1902 the old Kingdom of Zululand had been reduced to a rump, covering less than half of its original territory, which was similar in extent to the Colony of Natal. With much of its population living on land now purchased by whites, it was still home for about 200,000 people – approximately one-third of the African population in the old colony.[31] It was the exclusion of this recently annexed population that now threatened the workings of civil registration in Natal.

When Edward Dower, the former Cape Secretary of Native Affairs, began to act as the head of the new national department of Native Affairs he asked Arthur Shepstone – Theophilus' son and the Chief Native Commissioner of Natal – for advice about civil registration. Shepstone provided an enthusiastic assessment of the Natal system, concluding that, aside from problems of communication in the far northern districts of Ubombo, Ingwavuma and Hlabisa, he could 'see no reason why the births and deaths of all Natives should not be registered in Zululand; all that is required is that Act No 25 of 1902 be extended to that Territory and provisions made for the appointment of Official Witnesses'.[32] Much might have hinged on that mild word 'provisions', but Shepstone's suggestion was denied by an even more banal impediment. When Dower mentioned the idea of extending civil registration in to Zululand to the DOI he received a sharp rebuke from Smuts, reminding him that the NAD was not responsible for 'vital statistics in the Union' and demanding that 'any correspondence on the subject' should be transferred to his department. Having established which department owned this process, the Secretary for the Interior added that nothing was going to be done

---

[31] Aran S. MacKinnon, 'The Persistence of the Cattle Economy in Zululand, South Africa, 1900–50', *Canadian Journal of African Studies / Revue Canadienne Des Études Africaines* 33, no. 1 (1 January 1999): 101; John Lambert, 'From Independence to Rebellion: African Society in Crisis, 1880–1910', in *Natal and Zululand from Earliest Times to 1910: A New History*, ed. Andrew Duminy and Bill Guest (Pietermaritzburg: University of Natal, 1989), 386.

[32] Arthur Shepstone (Actg. Under Secretary for Native Affairs, Province of Natal) to Acting Secretary for Native Affairs, Pretoria, 'Registration of Births and Deaths (Natives in Zululand)', 29 May 1911, SAB NTS 9363 4/382 Natal and Zululand: Registration of births and deaths, 1911–1960.

for the foreseeable future.[33] In his report on the matter, Dower consoled his minister by reminding him that the 1907 commission had recommended that civil registration should be abandoned in Natal.

The most deadly blow to the system of civil registration in Natal was not administered by the Transvaaler Smuts, nor by one of the Cape magistrates; it came from within the provincial bureaucracy. W.J. Clarke was the Chief Commissioner of Police in Natal from the end of the nineteenth century. He was an ardent administrative progressive, and – as we have seen in Chapter 2 – a dogged champion of a plan for universal fingerprint registration.[34] In February 1913 he wrote to his national superior condemning the quality of the registration data being gathered in Natal. Clarke cited two cases of fraud in the Weenen district where Official Witnesses had been convicted of falsely recording births and deaths, in one of them 'as many as 60 or 70 false entries' had been recorded. 'It seems to me the system is open to fraud', he reported, 'and that the returns, as published, are very unreliable'.[35] Much later, after the system had been abandoned, the Natal magistrates would protest the injustice of Clarke's accusations, but his charges were received by R.H. Addison, one of the Zululand magistrates who had no experience or sympathy for the workings of the Official Witnesses. Addison all but admitted that fraud was rife, proposed that heads of households should travel with the Witnesses to confirm registrations, and then reminded the permanent secretary that the 1907 commission had, in any case, suggested that the whole scheme should be abandoned.[36] But even this unfriendly assessment was not sufficient to end registration; for a decade the Natal scheme carried on.

[33] *Ibid.*; Acting Secretary for Interior to Secretary for Native Affairs, Pretoria, 6 July 1911, SAB NTS 9363 4/382 Natal and Zululand: Registration of births and deaths, 1911–1960.; Edward Dower, [Acting?] Secretary for Native Affairs to Burton, Minister for Native Affairs, 'Registration of Births and Deaths of Natives in Zululand', 18 July 1911, SAB NTS 9363 4/382 Natal and Zululand: Registration of births and deaths, 1911–1960; Secretary for Interior, Pretoria to Secretary for Native Affairs, Pretoria, 10 March 1913, SAB NTS 9363 4/382 Natal and Zululand: Registration of births and deaths, 1911–1960.

[34] W.J. Clarke, Chief Commissioner, Natal Police, *Evidence to Natal Native Affairs Commission, 1906–7* (Pietermaritzburg: Government Printers, 1907); N.P. Pinto-Leite, 'The Finger-Prints', *The Nonqai* (1907): 31–3.

[35] W.J. Clarke, Chief Commissioner, Natal Police to Secretary, South African Police, 12 February 1913, SAB NTS 9363 4/382 Natal and Zululand: Registration of births and deaths, 1911–1960.

[36] R.H. Addison, Acting Chief Native Commissioner, Natal to Secretary for Native Affairs, Pretoria, 'System of Reporting Births and Deaths, Natal Natives', 24 February 1913, SAB NTS 9363 4/382 Natal and Zululand: Registration of births and deaths, 1911–1960.

## Retrenchment

The early 1920s were difficult times in South Africa, as all over the world, with the economy battered by inflation, fiscal retrenchment and working-class protests. It was amidst the very grim fiscal and political atmosphere of 1922 that the NAD found itself as the target of an efficiency evaluation by the independent committee of the Public Service Commission (PSC). The PSC took a scalpel to the administrative establishment that had been built up in the NAD after 1900, gutting, in particular, the regulators in the Johannesburg Government Native Labour Bureau and making the case that the whole of the department should be run along the lines of the skeletal administration in the rural districts of the Transvaal and Zululand.[37]

By October 1922 a special cabinet-level audit – appropriately named after the Geddes Axe Committee that had slashed government spending in Britain – had discovered the £2,000 that was being used to pay Official Witnesses in Natal. The new Chief Native Commissioner in Natal, C.A. Wheelwright, had himself come from the remote and backward Zoutpansberg district, and he looked askance at the civil registration scheme. When Wheelwright asked the Natal magistrates for their views on the abolition of the payments for Official Witnesses none of them agreed that it would be a good idea. All but one reported that it would be 'a source of great hardship' and mean the end of the system of civil registration.[38] It was Wheelwright who argued, on the basis of Clarke's 1913 police report, that the system was 'unreliable and not worth the expense involved' and, worse, that it encouraged fraud by offering a 'pecuniary inducement'.[39] Within weeks the payments had been suspended and the registration of African births and deaths immediately collapsed in Natal. The following year a new Births and Deaths Registration Act was published. The new Act, No. 17 of 1923, made registration of 'births, still-

---

[37] Dubow, *Racial Segregation and the Origins of Apartheid in South Africa*, 81–7; Saul Dubow, 'Holding "a Just Balance between White and Black": The Native Affairs Department in South Africa c.1920–33', *Journal of Southern African Studies* 12, no. 2 (April 1986): 217–39 shows this downsizing was not permanent, and by 1928, under the rich diet produced by the 1927 Native Administration Act, the NAD was back to its earlier weight. Falwasser, H.G., Acting Director of Native Labour, *Report of the Director of Native Labour* (Transvaal Administration, 1 November 1926), TAD GNLB 380 11/11 Native Affairs Department General, 1921–1935.

[38] Chief Native Commissioner, Pietermaritzburg to Secretary for Native Affairs, Pretoria, 'Registration of Native Births and Deaths', 22 December 1922, SAB NTS 9363 4/382 Natal and Zululand: Registration of births and deaths, 1911–1960.

[39] Secretary for Native Affairs, Pretoria, 'Memorandum: Geddes Axe Committee: Registration of Native Births and Deaths in Natal', 31 October 1922, SAB NTS 9363 4/382 Natal and Zululand: Registration of births and deaths, 1911–1960.

births and deaths of natives' compulsory only in urban areas, and left open the possibility that the DOI might declare registration compulsory in a specific rural district by regulation.

## The public health challenge

Over the next two decades the advocates of a state-supported public health system lobbied continuously for the restoration of registration in the countryside. These efforts peaked in the early 1930s when Natal and Zululand faced an unusually severe malaria epidemic. The champions of registration during this period were two local officials: Dr George Park Ross, at the time district health officer in the national Department of Health, and Harry Lugg, then the Native Magistrate for Verulam, a picturesque sugar milling town on the Natal north coast that was surrounded by malarial cane fields and impoverished native reserves. Both men had long experience of the skeletal government of indirect rule, an interest in the use of Shepstonian 'native agency' as a remedy, and, importantly, they would go on to take up the most senior provincial posts in their respective departments. Their combined efforts to restore the payment to Official Witnesses began in June 1932, when Lugg appealed to the permanent secretary of the NAD to restore the old system of registration, without which the severity of the malaria epidemic 'will never be known'. After reminding the permanent secretary that 'the last census did not include the enumeration of Natives' he urged the department to invoke the terms of the 1923 Act to require the 'registration of births and deaths of all natives'.[40]

A month later the most senior officials of the NAD and the newly formed Department of Health (DOH) met in Durban to discuss a combined response to the malaria crisis. The main resolution of this meeting, reflecting the ground work that Lugg and Park Ross had already completed, was that 'registration of births and deaths in Native Reserves should take place'.[41] While there were some signs of opposition, the principle, as the minutes show, was 'generally conceded' and the NAD accepted responsibility for the reintroduction of registration. Predictably, however, nothing happened. A month later, as the epidemic gathered steam, Park Ross wrote angrily to his departmental boss that 'nearly

---

[40] Harry Lugg, Native Commissioner, Verulam to Secretary for Native Affairs, Pretoria, 'Registration of Births and Deaths amongst Natives', 7 June 1932, SAB NTS 9363 4/382 Natal and Zululand: Registration of births and deaths, 1911–1960.

[41] E.H. Cluver, Asst Health Officer, 'Malaria Conference Held at Durban, Town Hall, on 1st July 1932', 4 July 1932, SAB NTS 9363 4/382 Natal and Zululand: Registration of births and deaths, 1911–1960.

every District in Natal is now affected with malaria, and until reliable information in the shape of an official registration of deaths is provided, it is impossible to deal effectively with outbreaks of the disease'. As the NAD magistrates attempted to estimate the number of deaths in the countryside from the epidemic, Park Ross railed against a policy that effectively made the large-scale deaths of black people in the countryside officially invisible. 'We have here an official disease', he protested, 'which, according to the returns has wiped out one-tenth of our population, and it apparently is to go unregistered.'[42]

Faced with the inertia of the NAD, the Health Secretary tried to goad his peers at the Department of Interior to remedy their glaring statistical shortcomings. This scientific reprimand pushed the Secretary of the Interior, after a delay of several months, to express his opposition to African registration directly. The registration of 'native births and deaths proved very unsatisfactory, both in Natal and in the Transkeian territories', he claimed, because 'Headmen appeared to have little interest in the duty'. The resulting registration of Africans was of 'small economic value', and the wisest policy (as far as the Department of Interior was concerned) was to suspend registration in the countryside until some future when 'the native may be sufficiently advanced in the scale of civilization to realise the advantages of registration'. The NAD, he concluded, was welcome to come up with a scheme for implementing registration, but it would not be acceptable to the statisticians (and accountants) at Interior unless it could be 'conclusively shewn that such a scheme would be of economic value and of real interest from a health point of view'.[43]

This cold rejection of the social value of African registration enraged Park Ross. He reminded his permanent secretary that 'an effective system of registering increase and deaths of cattle is in force in Native areas' and asked why it was 'impossible to evolve a similar system for human beings'. Park Ross acknowledged that there would be errors in the information provided by a system of registration in the Native Reserves but, contradicting the bleak view that the statistics of African births and deaths would be economically worthless, he insisted that the returns would be 'of the utmost value' because they would help to protect the public health and 'save numbers of lives'.[44] Park Ross promised to turn

[42] Chief Native Commissioner, Pietermaritzburg to Secretary for Native Affairs, Pretoria, 17 September 1932, SAB NTS 9363 4/382 Natal and Zululand: Registration of births and deaths, 1911–1960.

[43] H.J. de Wet, Secretary for Interior, Pretoria. Letter to E.N. Thornton, Secretary for Public Health, 9 February 1933. SAB NTS 9363 4/382 Natal and Zululand: Registration of births and deaths, 1911–1960.

[44] Senior Assistant Health Officer to Secretary for Public Health, 13 March 1933, SAB NTS 9363 4/382 Natal and Zululand: Registration of births and deaths, 1911–1960.

back to the NAD to try to get registration restarted. But the opposition of the Department of the Interior was sufficient to encourage J.M.Young, the Chief Native Commissioner (CNC) in Natal, to reject the idea. He declared that registration was worthless unless it could be implemented nationally, that the terrain and absence of roads in the reserves would make registration just another petty and unfulfillable burden and that Africans generally, and Chiefs and Official Witnesses in particular, were 'generally imbued with the same spirit of conservatism and distrust' that would undermine public participation. Shortly after Young had issued this rejection, he retired and was replaced as CNC by Harry Lugg.

Almost immediately Lugg tried to reopen the discussion about a special NAD registration effort. Drawing on his own long years as a magistrate he insisted that the claims that had been framed by the police and the officials at the DOI about widespread fraud were false. 'I have had considerable experience, extending over a period of twenty-one years, with the working of the Natal Act', he declared, and 'found no difficulty whatever in enforcing its provisions.' Against the official consensus that the individuals charged with responsibility for registration had generally neglected the work, he insisted that 'Native Official Witnesses carried out their duties faithfully and well'. And he contradicted the claim – circulating since the 1907 Native Affairs Commission – that Africans were generally suspicious of registration. 'The Natives regard the formal registration of a birth', he explained, 'as conferring added status to a child.' Finally, Lugg reminded his superior that the abolition of registration in 1923 was 'universally regretted by the Magistrates of this Province, for it left them with no means – except the Native Census (when held) – of ascertaining the increase of our Native population'. He now proposed a simpler and cheaper version of the original scheme, shifting responsibility for the one shilling payment for the registration of a birth onto the parents, but leaving the state responsibility for the fee for deaths. But Lugg's experience of the workings of the original system, and his insistence on its utility, could not soften his superior's concern about the financial difficulties the department faced in 1933.[45]

Park Ross and Lugg did not give up. They persisted throughout the 1930s in the effort to bring registration back to life. They were both very senior officials in their respective departments. But they faced entrenched inertia, skilled bureaucratic evasion and, most importantly, the unfounded but implacable view that African registrars would commit fraud to

---

[45] Harry Lugg, Chief Native Commissioner, Natal to Secretary for Native Affairs, Pretoria, 'Registration of Births and Deaths amongst Natives', 1 July 1933, SAB NTS 9363 4/382 Natal and Zululand: Registration of births and deaths, 1911–1960.

secure the payments for registration. By the middle of that decade, as the South African fiscal crisis dissolved away under the benign influence of an inflated gold price, the most senior figures in the NAD were using this argument – derived solely from Clarke's 1913 report in the official record – to overturn Park Ross' claims about the public health value of registration. Here they invoked Clarke's claim that registration events would be invented for gain which would mean that 'our vital statistics, about which the doctors are making such a fuss, will all go wrong'.[46]

As this last comment suggests, the effort to restore registration in South Africa was not the isolated obsession of Lugg and Park Ross. Throughout this period groups of doctors kept pushing for compulsory vital registration in the countryside. In their efforts to rationalise the missing data, the leading officials in both the NAD and the Interior presented explanations that would have amused Franz Kafka. In 1942, for example, while the Allied armies were still on the defensive around the world, the hearings of the Beveridge Commission in Britain prompted the South African government to plan seriously for a national health service that would include the African population in the countryside.[47] The division of the South African Medical Association (SAMA) for the Transkei, based as they were in the middle of the largest native reserve, made a public call for the restoration of civil registration to strengthen the political case for a nationalised health service. They wrote to the Secretary of Native Affairs, reminding him that their organisation was 'wholeheartedly in favour of a State Medical Service for Natives' and asking the NAD to implement civil registration. Here again the doctors drew out the embarrassing comparison between the state's expectations for livestock and the requirements for people: 'If it is found necessary to register the Births and Deaths of cattle, it would appear that the registration of the Births and Deaths of the people should be even more necessary.' They cited the data from a single hospital in the Transkei to show the horrible effects of tuberculosis in the countryside – over 20 per cent of total deaths were derived from this single cause. 'Reliable statistics', they argued in a statement that echoes through to the recent HIV crisis, 'would make the need for a [national health] service abundantly clear.'[48]

[46] D.L. Smit, Secretary for Native Affairs, Cape Town to Harry Lugg, Chief Native Commissioner, Natal, 'Registration of Births and Deaths of Natives', 26 February 1935, SAB NTS 9363 4/382 Natal and Zululand: Registration of births and deaths, 1911–1960.

[47] Shula Marks, 'South Africa's Early Experiment in Social Medicine: Its Pioneers and Politics', *American Journal of Public Health* 87, no. 3 (1 March 1997): 452–5; Seekings, 'Visions, Hopes & Views about the Future', provides a very useful explanation of this period.

[48] Arnold H. Tonkin, Hon Sec, South African Medical Association, Border Branch, Transkei Division. Letter to H. Lawrence, Minister of Interior. 'Registration of Births and Deaths

The NAD's response to this request highlights the ignominious deferment that had by this time become habitual in the department's responses to this question. 'What would be the use of statistics when the causes of death are inaccurate?' The official responsible for registration replied that 'the state med[ical] service must come first, so that all Natives can have med[ical] attention. After that we can get accurate statistics'.[49] The champion of the South African national health service, Henry Gluckman, was appointed Secretary of Health in 1945, but that was not enough to deliver his proposals. After 1943, as the Allies' global position improved, the bureaucracy's enthusiasm for social interventions declined rapidly, replaced by an increasingly feverish anti-communism and a determination to use force to stifle dissent among the black population.[50] In the debates that followed, the extent and effects of disease in the countryside remained largely invisible, a statistical situation that has only begun to change very recently.[51]

In the years immediately after the Second World War organisations like SAMA, the National Child Welfare Association, the Anti-Tuberculosis Association and the African representatives in the Ciskeian General Council placed almost continuous pressure on the state to act.[52] In 1946 the NAD, 'being pressed on all sides to secure the introduction of a system of registration at the earliest possible date', attempted, again without any success, to persuade the Department of Interior to take up the task.[53] Over the next few years they began to plan, fitfully, for a return to the system of Official Witnesses that had been working before 1922. Finding resources for registration remained a problem, even as the pensions for Africans were being paid out. In the new registration plan the fee to be paid to African registrars had been whittled down to three pence – in real

in Rural Native Areas', 28 April 1942. SAB NTS 9360 2/382 Part 2 Registration of Births and Deaths (natives) in the Union. (general File). 1947 to 1952.

[49] See marginalia on Arnold H. Tonkin, Hon Sec, South African Medical Association, Border Branch, Transkei Division to H. Lawrence, Minister of Interior, 'Registration of Births and Deaths in Rural Native Areas', 28 April 1942, SAB NTS 9360 2/382 Part 2 Registration of Births and Deaths (natives) in the Union. (general File). 1947 to 1952.

[50] T. Dunbar Moodie, 'The South African State and Industrial Conflict in the 1940s', The International Journal of African Historical Studies 21, no. 1 (1 January 1988): 21–61.

[51] Hoosen Coovadia, Rachel Jewkes, Peter Barron, David Sanders and Diane McIntyre, 'The Health and Health System of South Africa: Historical Roots of Current Public Health Challenges', The Lancet 374, no. 9692 (5 September 2009): 817–34.

[52] Secretary for Native Affairs, 'SAB NTS 9360 2/382 Part 2 Registration of Births and Deaths (natives) in the Union. (general File). 1947 to 1952', 1952 1947, Central Archives, Pretoria, www.national.archsrch.gov.za/sm300cv/smws/sm30ddf0?20080630 2009567CA78806&DN=00000002.

[53] Mears, Secretary for Native Affairs to Secretary for the Interior, Pretoria, 'Registration of Births and Deaths in Rural Native Areas', 10 May 1946, SAB NTS 9360 2/382 Part 2 Registration of Births and Deaths (natives) in the Union. (general File). 1947 to 1952.

terms a tiny fraction of the one-shilling payment that Official Witnesses had earned half a century earlier. The total that the NAD planned to spend on national registration – an additional amount of £5,000 per annum – was never a very large sum and it was certainly well within the state's reach throughout this period.[54] But even these desultory efforts were swept away by the National Party's plans for Native Affairs after 1948. As the department turned to building elaborate and compulsory labour bureaus for African workers in the cities, and then, after 1952, to the even more ambitious and costly system of reference books for all black adults, urban and rural, the prospects of a national project of civil registration dimmed.[55]

By the end of the 1940s, the arguments about the virtues and prospects of civil registration had come full circle. Almost half a century earlier W.J. Clarke had suggested that registration, in the absence of fingerprinting, was 'open to fraud' and unreliable. For decades that flimsy (and fiercely contested) characterisation of African registrars as frauds had been used as a pretext by those who sought to resist the continuous calls for a reliable system of civil registration. As the Apartheid state began to take form, with (as we will see in the next chapter) a very large and renewed interest in fingerprinting as a remedy to the limits of the official documents of identification, the most senior officials in the NAD, including those who had earlier been part of the Lugg–Park Ross campaign to restore registration, began to argue that only biometric registration could reliably establish the identity of Africans in the countryside. 'In view of the vast illiterate population, many of whom have no address or fixed abode, the Department is of opinion that the only practicable means of registration', Frank Rodseth, the departmental Under-Secretary, wrote, 'is by a system of finger prints.'[56]

## Conclusions

On the face of it the story of the failure of civil registration in South Africa is a familiar one of colonial state-building on the cheap.[57] The officials

---

[54] H.H.L. Smuts to Secretary for Native Affairs, 'Memo', 10 March 1949, SAB NTS 9360 2/382 Part 2 Registration of Births and Deaths (natives) in the Union. (general File). 1947 to 1952.

[55] Douglas Hindson, 'Pass Offices and Labour Bureaus: An Analysis of the Development of a National System of Pass Controls in South Africa' (PhD, University of Sussex, 1981), 184–212.

[56] F.R. for Secretary for Native Affairs to Director of Census and Statistics, Pretoria, 'National Register of Population'.

[57] Walter Rodney, *How Europe Underdeveloped Africa* (Washington: Howard University Press, 1982), 200–30; Frederick Cooper, *Decolonization and African Society: The Labor*

charged with making registration work simply lacked the funds to do it. It is certainly the case that the Native Affairs Department – without a constituency that was represented in parliament and often despised by the most influential African leaders – was always on the defensive about its budgets. Yet the amounts involved in building a large system of registration were actually very small. Even more importantly the advocates for registration typically came up with innovative ways to make registration pay for itself, and they often preferred to use existing instruments of indirect rule (like headmen and dipping inspectors) to achieve it. Neither argument succeeded in overcoming the scepticism or vacillation of the bureaucracy. This history suggests that the South African state had the resources and the ambitions to expand from its position as gatekeeper; the fact that it did not do so was – as many critics observed as it was happening – a product of its very limited ability to gather information about people in the countryside. In this sense the South African gatekeeper state – concerned to police and administer Africans only when they were in the cities – was a result, and not the cause, of the failure of universal registration. Nor was this disinterest in the details of the lives of rural black people pre-ordained by the system of indirect rule; the reverse might easily have been true.

This brings me to the second straightforward explanation for the failure of registration: racism and racist views. Some of this is certainly correct. It was clearly the case that racist contempt informed the idea that all African registrars would act as frauds. The persistent failure of any real urgency on the part of the officials (especially at the DOI) charged with making government work in the tribal areas clearly followed from the absence of any meaningful sympathy with the black people who lived there. In this respect registration in South Africa follows a pattern of racialised government that is familiar elsewhere in Africa and the United States.[58] Nor can there be much doubt that the absence of reliable statistics on the appalling health conditions amongst black people in the

*Question in French and British Africa* (Cambridge University Press, 1996), 465–6; Frederick Cooper, *Africa Since 1940: The Past of the Present* (Cambridge University Press, 2002), 5–6; M. Crawford Young, *The African Colonial State in Comparative Perspective* (New Haven: Yale University Press, 1994), 120; Berry, 'Hegemony on a Shoestring'.

[58] Theda Skocpol, *Protecting Soldiers and Mothers: The Political Origins of Social Policy in the United States* (Cambridge, MA: Harvard University Press, 1992), 533; S. Shapiro, 'Development of Birth Registration and Birth Statistics in the United States', *Population Studies* 4, no. 1 (1950): 98–9; Fred Cooper, 'Voting, Welfare, and Registration: The Strange Fate of the Etat-Civil in French Africa, 1945–1960', in *Registration and Recognition: Documenting the Person in World History*, ed. Keith Breckenridge and Simon Szreter, Proceedings of the British Academy 182 (Oxford University Press, 2012).

countryside helped to strengthen the policy of official neglect. But racism cannot sufficiently account for the complete and prolonged failure of civil registration, not least because the most successful periods of registration coincided with the eras of the fiercest white supremacy.[59]

A more convincing explanation emerges from the forms of legal and political representation in South Africa (and elsewhere on the continent). Repeatedly a well-worked out scientific argument about the costs and benefits of civil registration was allowed to die in the disagreements between departments, or amidst the more urgent fiscal demands of the Depression or the World War. It was the absence of a political constituency arguing for registration that allowed this inertia. Here the contrast with the many different forms of registration in other regions of the world is striking, because both elites and the poor – who, in other contexts, sought registration for many different ends – were silenced by the workings of customary law.[60] Indirect rule in rural South Africa, as in most of Africa, placed administrative processes in the hand of chiefs who were charged to substitute for a developing public sphere and political economy of literacy that was occurring in the towns.[61] Customary law ensured that the legal arrangements for property typically avoided registration in the countryside.[62] Even more importantly, literate religious elites, which played very important roles in Europe, the Americas, China and Japan (and had been key agents of social and cultural transformation in the nineteenth century in Southern Africa) were denied authority over people and property, probably because their loyalties were suspect from both sides.[63] In the absence of either a popular or an elite constituency advocating for registration it was always an easy option for the bureaucracy to delay. This history of the colonial state suggests that, as Stoler has shown, an overarching imperial 'homage to reason was neither pervasive nor persuasive'.[64]

[59] See especially John W. Cell, *The Highest Stage of White Supremacy: The Origins of Segregation in South Africa and the American South* (Cambridge University Press, 1982); M. Lake and H. Reynolds, *Drawing the Global Colour Line: White Men's Countries and the International Challenge of Racial Equality* (Cambridge University Press, 2008).

[60] Keith Breckenridge and Simon Szreter, eds, *Registration and Recognition: Documenting the Person in World History*, Proceedings of the British Academy 182 (Oxford University Press, 2012).

[61] See Ntsebeza, *Democracy Compromised*, for an excellent account of how this worked in South Africa.

[62] Sara Berry, *No Condition Is Permanent: The Social Dynamics of Agrarian Change in Sub-Saharan Africa* (Madison: University of Wisconsin Press, 1993).

[63] Comaroff and Comaroff, *Of Revelation and Revolution*, vol. 1, 261–87, 305; see the essays in Breckenridge and Szreter, *Registration and Recognition*.

[64] Ann Laura Stoler, *Along the Archival Grain: Epistemic Anxieties and Colonial Common Sense* (Princeton University Press, 2008), 58.

But there is another intriguing irony at work in this story. As we will see next, from the early 1950s the South African state undertook a massive project of fingerprint registration. It was chaotic and expensive and famously resisted by individuals and organisations. Yet, as I show in the last chapter, by the time of the establishment of full democracy and the beginnings of a universal system of social benefits, the entire adult population of South Africa had been registered biometrically. And it was this centralised biometric population register which made possible the rapid and very wide delivery of the cash payments for child support and pensions that are being acclaimed around the world; it is this extremely efficient and highly scalable technology which underpins the current interest in a global system of cash grants for the poor.[65] It is especially paradoxical, in light of the fraught history of South Africa, that the possibility for the worldwide unconditional social assistance regimes emerged from the state's preoccupation with fingerprinting as an alternative to local civil registration.

[65] J. Hanlon, D. Hulme and A. Barrientos, *Just Give Money to the Poor: The Development Revolution from the Global South* (Sterling: Kumarian Press, 2010), 38–9; J. Hanlon, 'It Is Possible to Just Give Money to the Poor', in *Catalysing Development? A Debate on Aid* (Oxford: Blackwell, 2004), 199.

# 5    Verwoerd's bureau of proof: the Apartheid Bewysburo and the end of documentary government

On 10 March 1952, A.J. Turton, the Senior Urban Areas Commissioner of the South African Department of Native Affairs, received a very urgent telegram from Werner Eiselen, the Departmental Secretary, instructing him to take the next train to Cape Town. After completing the eighteen-hour journey from Pretoria, Turton was met by a government driver at the Cape Town station and taken to meet H.F. Verwoerd, the new Minister of Native Affairs, future prime minister, and the architect of Apartheid. In meetings with Verwoerd, Eiselen and the state's legal advisers over the next few days Turton set down one of the legal pillars of Apartheid – the Abolition of Passes and Co-ordination of Passes Act. This infamous law laid down the regulatory basis for a new form of personal identification and the bureaucratic mechanism to enforce the nationwide registration of all Africans, their names, locale, tax status, fingerprints and their officially prescribed rights to live and work in the towns and cities of South Africa.

The document issued to all Africans in the course of the 1950s and 1960s was called, in English, the Reference Book. All the details of the book, and any changes to the status of its bearer, were to be recorded and managed from a single government office, the Central Reference Bureau in Pretoria. The political, and emotional, investment that the bureaucrats made in this development is better captured by the Afrikaans terms, Bewysboek and Bewysburo – the book and bureau of proof. Africans had another name for the Bewysboek. They called it the Dompas – the stupid pass – and that name captures it best. For the Dompas, more than anything else, defined the essence of life in Apartheid South Africa. It was intended to sweep away the unreliable documentary order that had dominated South Africa in the years after 1910. In this it succeeded. As we will see, the new 'Book' erased the politics of writing associated with the archival state of the early twentieth century. It was 'dom' in both senses of the Afrikaans word – dumb, in

An earlier version of this chapter was published as 'Verwoerd's Bureau of Proof: Total Information in the Making of Apartheid'. *History Workshop Journal* 59 (2005): 83–109.

that it curtailed writing, and stupid – because it was broken. In almost every respect, as we shall see here, the Dompas was a complete failure, but the form of its failure defined the character of the Apartheid state, and much of its democratic successor.[1]

After 1960 the Dompas also became the outstanding global metaphor of violent white supremacy. The shocking events of 21 March 1960, in which the police opened fire on a large crowd of anti-pass protesters in the model township of Sharpeville, made South Africa – and the protests against passes – priority news on the world's television networks and newspapers. The graphic violence directed at the protesters played directly into protest movements in Europe and the United States that had been developing during the 1950s.[2] After Sharpeville, South Africa and the Dompas in particular became metonymic for white supremacy everywhere, sustained by a distinct iconography of abjection with powerful images produced by local and international photographers of white policemen inspecting passes, of work-seekers queuing dolefully to secure their documents, and of violent police raids.[3]

## Blueprint for the Dompas

A.J. Turton was a veteran English-speaking bureaucrat, with little in common with the Nationalist Party newcomers, Eiselen and Verwoerd. He had worked his way up through the ranks of the Native Affairs Department after joining in 1928. He had been summoned to Cape Town because of a memorandum he had drawn up on 17 October 1950 – just two days after Verwoerd had replaced E.G. Jansen as Minister of Native Affairs. This document gave detailed substance to the old ideal of the universal, centralised registration and control of all African adults. As we have seen, this panoptic fantasy had a long life dating back to Edward Henry's brief tenure on the Witwatersrand in 1901. Turton's specific plan for the Central Reference Bureau and the radical simplification of the regulations governing influx control had been developing within Native Affairs since the end of the 1930s. It was an old plan, designed to reconfigure the contested governance of the previous decades, and to support a set of new Apartheid

---

[1] CAD NTS 9791 1004/400, Secretary of Native Affairs to U S Admin, Telegram TC 276/278, 10 March 1952.
[2] Christabel Gurney, '"A Great Cause": The Origins of the Anti-Apartheid Movement, June 1959–March 1960', *Journal of Southern African Studies* 26, no. 1 (1 March 2000): 123–44; Tom Lodge, *Sharpeville: An Apartheid Massacre and Its Consequences* (Oxford and New York: Oxford University Press, 2011), 228, 234–79.
[3] Manning Marable, *Malcolm X: A Life of Reinvention* (New York: Viking, 2011), chap. 8, 11; see Martin Luther King lobbying for sanctions at the White House in 1962, Lodge, *Sharpeville*, 228.

mechanisms, in particular the Labour Bureaus and the new racially defined Group Areas. As Turton put it, the passing of the Population Registration Act, no. 30 of 1950, marked 'a great opportunity to do some thinking on entirely new lines and to break away as completely as possible from the outworn portion of our existing pass laws, tax and other procedures'.

Turton proposed to sweep away all the old contracts, tax receipts and paper passes, and to replace them with just two official documents, an identity card and something he called a 'personal file'. This booklet was to contain the 'personal history and movements' of every African worker, and a long list of official permissions which, up to that date, had been granted on separate pieces of paper. These included official permission to enter an urban area, permission to seek work, records of required medical examinations and any particular medical history, the names and addresses of employers, and, most importantly, receipts for tax payments. Appealing to the new minister's preoccupation with utilitarian cost-saving, Turton proposed to distil the entire interaction between Africans and the state into this single 'personal file', massively reducing the opportunities for impersonation, and the cost to the state in a single step.[4] 'A great saving of staff would be made', he pointed out, 'as two million identity documents and tax receipts are issued every year; with each succeeding document becoming less a proof of identity and more a scrap of useless paper.' In the process he proposed the simplification of all other tax and pass legislation, replacing local and provincial regulations with a single focus on all towns and cities regulated by the Labour Bureaus, and hinging permission to seek work on the record of tax payments.

Coordinating all of this required the establishment of a new Central Identity Bureau, which would be responsible for maintaining the integrity of the identity cards, and, Turton casually added, collecting 'reports of desertions, tax payments and even perhaps the District where the Native is employed'. By framing his proposal around the cost-efficiency of a central registration system, Turton put together a devastatingly persuasive argument for a panoptic mechanism of identification, taxation and policing. Buried deep in the bowels of his proposal was the suggestion that the central bureau would provide a mechanism for the complete control of the African population. Requests from the police, tax and municipal authorities for 'the tracing of offenders, tax defaulters, and honest people whose whereabouts are urgently sought' would not only be handled by the central bureau, but so efficiently that it would, once

---

[4] For more on Verwoerd's administrative preoccupations see Ivan Evans, *Bureaucracy and Race: Native Administration in South Africa* (Berkeley: University of California Press, 1997), 62–73.

again appealing to the parsimony of the new minister, 'cut down the cost of administration in those respects by a startling percentage'.

In two respects Turton's project marked an abrupt ideological and administrative break with the earlier practices. First, the new system of registration, which left no space for African workers, or even their employers, to write on the 'personal file', meant the death knell of the ideology of the mutually agreed, signed and witnessed contract, which, as we have seen, had underwritten both the role of the Native Affairs Department and the labour law before Apartheid. Remarking that abuses of written contracts of service were 'so rare as not to justify the further continuance of an outmoded and costly system', Turton blithely dismissed the ideological mantra of the early twentieth-century officials. And he naively ignored the realities of employment and policing throughout the country, suggesting that an abusive employer would simply 'lose all his servants'.

In the second area, Turton ushered into being, almost by accident, the project of universal fingerprinting that had been proposed, and rejected, in every previous decade. The scepticism nurtured by Gandhi's protests and sustained by the officials responsible for the existing large fingerprint collections on the mines and in the police was all but erased by the allure of the cost-cutting efficiency of centralised registration. Turton presented fingerprinting as a means to the end of extraordinary efficiency, not, as had been the case in earlier arguments, a tool for policing. 'Finger prints will have to be taken when issuing identity cards', Turton noted in the last paragraph of his proposal, 'otherwise as soon as his card or personal file become inconvenient a native could destroy them and obtain a new identity card with a new serial number and so defeat the whole purpose of the identity card system.' That the entire apparatus hinged on the efficacy of the fingerprinting systems went unnoticed in the early phases of the Dompas campaign.[5]

## Implementing the plan

Well before Verwoerd's appointment as minister, the department had been looking for ways to simplify the old tax and pass regulations, and implement the new Apartheid policies. These measures were particularly focused on the new racially-defined Group Areas and the Labour Bureaus. In October 1951, Verwoerd convened a committee of members of the Departments of Justice, Police, Census and Native Affairs

---

[5] CAD NTS 9791 1004/400, Turton, A.J. Senior Urban Areas Commissioner to Secretary of Native Affairs, 'Identity Cards: Alteration of Pass Laws; Registration and Permit System', 17 October 1950.

to devise a mechanism to integrate the pass system and the Population Registration Act, to find a way to simplify the pass regulations and to reduce the twenty-eight different documents and forms Africans were required to complete and carry. Turton's plan provided exactly what the committee, and Verwoerd, were looking for. It was, unsurprisingly, accepted as a catch-all solution to the problems that the state faced in all these areas – with just one prerequisite improvement to Turton's plan. A mechanism, the Inter-Departmental Committee resolved, needed to be found that would allow for the lamination of the identity card, physically preventing any possibility of forgery.

Early in January 1952, the South African Bureau of Standards (SABS) was given the task of establishing whether a government document could be manufactured that would preclude further written modification. They evaluated two of the new post-war plastic materials – Jewellex Adhesive Cement and Durex Self Adhesive Tape – against the most popular methods for altering official documents. In the SABS tests the materials withstood submersion in boiling water, heat treatments and steaming. And when subjected to a variety of chemical solvents, the paper and its bonded plastic cover dissolved together. So for the first time the state could look forward to the possibility that the documents it issued would be protected from any further writing.

Immediately after the successful lamination report was submitted, Verwoerd called Turton to Cape Town to begin drafting the Native (Abolition of Passes and Co-ordination of Documents) Bill. The new law swept away the old forms of paper-based personal identification drawn from tax receipts, and service contract numbers, each of which, through duplication or impersonation, led to 'endless confusion' in the Labour Bureaus and revenue offices of the NAD. In its place the law inscribed the reference book number as the key index of personal identification for black people. (It is worth noting that this was not, yet, the ID number South Africans are familiar with today. That innovation emerged only later after a campaign waged by officials in the Reference Bureau around the slogan, 'een naturel, een nommer – one native, one number'.) Highlighting the conflicting and draconian pass requirements from the colonial period, the law abolished legislation that had applied to the individual provinces and cities, leaving in its place a radically simplified requirement based on the Labour Bureau system set up by the Native Laws Amendment Act passed in 1949 and amended in the same year. Under these regulations Africans from the countryside required permission from their local magistrate or Native Commissioner to travel to any urban area, and once there they required permission to work from the local Labour Bureau. No employer was permitted to offer work to any African who had failed to register or secure work permission from the Labour Bureau. And the primary task of the Labour Bureau was to limit and, eventually, reduce the number of Africans present in the cities.[6] All of these administrative procedures were to be recorded in the new Reference Books, and the books were to be harnessed to a laminated identity card that every adult, regardless of race, was required to carry.

## Issuing the books

The process of issuing the Reference Books began in earnest in March 1953. Using a production line technique that they called the 'factory system', mobile teams of NAD officials visited the large government and industrial employers around Pretoria, the Witwatersrand and the Vaal Triangle over the next six months. The Chamber of Mines initially refused to allow workers to be registered at individual mines or as they arrived at the WNLA's Mzilikazi Compound on Eloff Street Extension; they were determined to wait until registration had begun in the Transkei before allowing the Reference Books, and the potential protests they might precipitate, into the compounds. Other employers were not as squeamish. 'Die reaksie van die sakeondernemings was werklik 'n ontnugtering', Eiselen reported to Verwoerd from his observations of the registration process around Pretoria, 'Geen besware en die volste samewerking is ondervind' ('The reaction of the businesses was really sobering. No objections and the fullest cooperation were experienced'). In some cases businesses paid the fees charged for photographs. At the Olifantsfontein Brick Works, Lady Cullinan herself arrived to be photographed along with the workers.[7]

By the end of October, the mobile teams had completed the work of issuing books in the mid-sized industrial towns of Springs, Benoni, Boksburg, Germiston, Brakpan, Kempton Park, Alberton and Brits. Teams were still at work in Johannesburg, Roodepoort, Randfondtein, Witbank and Heidelberg. In the month of July 1953, the twelve teams in the field registered 80,000 workers, a figure that represented the apex of the registration process. The heart of the non-mining industrial

---

[6] CAD NTS 9794 1031/400 Discussions between Senior Officers Head Office 1951–4. 'Instelling van persoonbewyse vir die Bantoe', no date and Van Heerden 'Bespreking oor Arbeidsburos', 21 October 1952.
[7] CAD NTS 9793, 1027/400(9) Matters referred to U/S (E.A.) Central Reference Bureau. Eiselen to Verwoerd, 6 February 1954.

proletariat, some 400,000 workers, had been successfully issued with 'bewysboekies' by the end of October.[8]

One success followed another. The police loved the new system because it freed them from having to understand and investigate the many different forms of written authorisation that had characterised the old documentary order. And, more importantly, it gave them a mechanism that allowed for policing without the requirement for conversation. 'The investigation of the Natives does not waste time because of the presentation of the wrong documents', L.J. Lemmer, Chief Clerk of the Bewysburo explained, 'and the [need for a] series of questions also falls away. The police official just asks for the bewysboek and simply looks through it.'[9]

Then, as now, it was the promise of a massive simplification of policing practice, and a vast, asymmetrical, expansion of the powers of the police, that lay behind the effort to implement a mechanised documentary panopticon. In the same month that the registrations reached their highest point Turton and the South African Police began to set in place procedures which would allow for the use of the Bewysburo as a centralised, universal, detection facility. After noting that the 'fingerprints of every native male over 16 years of age' were to be taken, classified and associated with a number that 'he will retain under all circumstances', the Deputy Commissioner of the South African Police described the workings of the Bewysburo as the modern, bureaucratic, equivalent of Bentham's panopticon. 'A card is prepared in respect of each native and properly filed in a filing cabinet accommodating 60,000 cards.' Looming over these cards, each cabinet would be controlled by 'a female clerk' who would have, at her fingertips, the identity and location of 60,000 individuals.

The new system promised to allow the police swiftly to identify and locate any 'Native whose number is known, or whose fingerprints are available'. After applying, first, at the offices of the local Native Commissioner or municipal Registration Officer, 'the number and/or fingerprints could then be forwarded to Pretoria where the various numbers are all to be numerically compiled into lists ranging from 1 to 60,000, 60,000 to 120,000, etc. Two copies of the lists are then to be submitted to the Bureau once or twice fortnightly for submission to the clerks in charge of the various cabinets.' Here, at last, was a single, cost-effective tool that would allow the police to track elusive African suspects. The Commissioner

---

[8] CAD NTS 9794, 1031/400 Discussions between Senior Officers Head Office. 1951–4. Lemmer, L.J. 'Naturelle (Afskaffing van Passe em Koordineering van Dokumente) Wet, 1952', 26 October 1953, and 'Samespreking tussen Senior Aptenare', 26 October 1953.
[9] *Ibid.* (Lemmer).

issued instructions for police throughout the Union to begin compiling lists of 'Native suspects' for processing by the Bewysburo.[10]

The new system seemed to offer similar solutions to the previously chaotic and expensive tax collecting division of the Department of Native Affairs. The first results were very promising. As an immediate and direct consequence of the issuing of the Dompas, the amount of tax collected from individual workers began to increase dramatically. Workers could not complete their registration (and hence secure their continued right to work) without first paying up their outstanding taxes. In the districts where the books had been issued the amount collected in the months from March to October, 1953, was some £300,000 greater than the sum collected in the previous year. A collection of the senior officers of the NAD marvelled at an oral report of the workings of the Bewysburo in October 1953, with Eiselen concluding that 'the cost of the whole operation has already been more than recovered by the additional tax received, and in addition to the fact that we have achieved our primary goal with the operation, there is a bumper harvest of associated benefits, so that we are getting a good and sound mechanism without any cost'.[11]

Even at this early stage there were some pertinent objections, of course, but they are significant primarily for the manner in which they were dismissed. The liberal Durban city Native Affairs Manager, Eric Havemann, characteristically queried the sudden death of the written contract, arguing in the precise terms used by his predecessors that it was essential for the 'protection of the natives' interests'. Unlike Turton, and the others in Pretoria, Havemann had a close understanding of the persistent conflict between African workers and their employers, and of the utility of the written contract in resolving these conflicts. 'This office deals with between 200 and 300 complaints per month from Natives alleging breach of contract on the part of their employers', he wrote to his local Commissioner in July 1953, 'a task rendered straightforward by the fact that all contracts of service are registered only if the Native agrees with the terms thereof.'[12] Under the new system there was no possibility of any official sanction of the verbal contracts entered into between white employers and their African servants, and not even the pretence that consent on the part of African workers was of any

[10] CAD SAP 494, 15/2/52, 'Uitreiking van Bewysboekies aan Bantoes, 1953–7'. Deputy Commissioner, Transvaal Division SAP to Commissioner of the SAP. 29 July 1953.
[11] CAD NTS 9794, 1031/400 Discussions between Senior Officers Head Office. 1951–4. 'Samespreking tussen Senior Aptenare', 26 October 1953.
[12] CAD NTS 1027/400 Naturelle Sake. Bewysburo (b), Havemann, E. to Native Commissioner, Durban, 'Department of Native Affairs: Consolidated Standing Circular Instructions re Population Registrations of Natives', 27 July 1953.

consequence. In his response on Turton's behalf, Lemmer made clear the newly diminished status of contract in the technocracy of Apartheid. The key ideological function of the contract – that it gave Africans the (illusionary) choice of the terms of their labour and allowed the state to act as the arbiter of this consent – was 'outweighed by the benefits such as the saving in man hours of labour, the saving of inconvenience to employers and employees, and the reduction in the number of documents to be carried by the Native'. Lemmer supplemented this bureaucratic justification with another, as manifestly dishonest as the former was crudely pragmatic. 'The exclusion of the particulars of service contracts from reference books was the result of the desire expressed by the Natives consulted in the Chief Native Commissioners' areas throughout the union', he lied, 'that such particulars should not be included as their inclusion would diminish the Native's bargaining powers in regard to wages.' Turton's original proposal, you will remember, had included a conscious dismissal of the continuing significance of the contract. And his rejection of the legal cornerstone of the segregationist period was, undoubtedly, one of the reasons why his plan was so attractive to Verwoerd.[13]

What were the implications of the new regime for African workers? This is a question that hangs over this entire discussion, and it is not easy to answer briefly. There were some immediate effects. The buildings that combined the offices of the Labour Bureau and Influx Control officials in all towns became the pivot of working life in the Apartheid period. It was to these offices, usually the Arbeidburo, that all workers, including women for the first time after 1958, had to report on entering an urban area, and it was in here that workers received permission to seek work (or not). It was in the Arbeidburo that workers encountered the medical regulations and humiliating inspections required by particular local regulations. And it was the Arbeidburo that served as the pivot for the registration of work. Regardless of the specific political interventions of individual managers, the general force of the law, and policing, made all workers subject to a massive new disadvantage in relation to their employers. This was especially true for workers bound to work on the farms at the discretion of the landlord by the words 'Farm Labour Tenant' stamped into the Dompas. But in most respects it was not the working effects of the Bewysburo that mattered for individual workers.

---

[13]  CAD NTS 1027/400 Naturelle Sake. Bewysburo (b), Lemmer, L.J. for Director, Central Reference Bureau to Chief Native Commissioner, Pietermaritzburg. 'Native Affairs Code- Population Registration of Natives', 16 November 1953.

The most appalling consequences, and the full answer to this question, require us to examine the effects of its failure.

## Problems

The first signs of trouble in the coordination of the Bewysburo came just weeks into the programme of delivery. In the first document to lay out the manpower requirements of the fingerprinting scheme, L.J. Lemmer presented some basic arithmetic to his superiors in June 1952. Given the 'recognised fact that no person can [reliably] complete more than 30 fingerprints per day', he estimated that the new Bewysburo would require 66 'finger print experts' just to keep up with the natural growth in the pool of sixteen year olds. An additional group of at least ten experts would be required for 'court evidence, the investigation of deserters, false documents, and impersonation'. Aside from mentioning that there would be a two-year lag between the registration and the processing of finger-prints, and offering some ominous warnings about the rate of processing currently under way at the Native Affairs Finger Impressions Record Department, Lemmer was remarkably silent about the most important task ahead of them, the processing of the fingerprints of the 2.5 million adult African men (and a similar number of women) who were to be issued Reference Books. His superiors – Turton, Eiselen and Verwoerd – seem not to have paid any attention to the arithmetic implications of Lemmer's report.[14] Nor did they consider the particular staffing problems associated with finding, and retaining, fingerprint experts.

One of the ironies of the early Apartheid period is that the massive reawakening of the state's interest in Native Affairs actually weakened its ability to act. As the new policies began to orient much of South African municipal and industrial life around the complex new set of segregation-ist laws and regulations, so the NAD had increasing difficulty finding and retaining suitable candidates for its burgeoning bureaucracy.[15] The new Bewysburo was plagued by staff shortages almost as soon as the registration campaign got under way. By the start of 1954 Turton had £500,000 in increased taxes to his credit as he started to lobby the Public Service Commission for five additional mobile teams to begin the task of distributing the books in Natal, and for twenty additional clerical work-ers to process the work of the new teams.[16] 'I am desperate', he wrote

---

[14] CAD NTS 1027/400 Naturelle Sake. Bewysburo (b). Lemmer, L.J. 'Native Affairs Memo: Sentrale Vingerafdrukburo', 10 June 1952.

[15] Evans, *Bureaucracy and Race*, 73–89.

[16] CAD NTS 9794, 1027/400(13) Central Reference Bureau. Forging of Reference Books. Director, Central Reference Bureau to Secretary of Native Affairs. 18 January 1954.

to a friend in the Ministerial offices in March, 'I have had wonderful co-operation from ... Govt, Urban Areas, Brand, etc but what staff branch has [done] to this Bureau will make your hair stand on end'. And he appealed for an opportunity to state his case in person to the minister.[17] Having built a case for the Bewysburo on the basis of its parsimony, he now found himself in the impossible position of having constantly to defend his budgeted complement, and lobby for its expansion. As the staffing crisis continued Turton was to be hoist by his own cost-cutting petard.

On the first anniversary of the Dompas campaign, with well under a million books issued, the first signs of very serious trouble began to appear. Turton's immediate supervisor, C.A. Heald (the Undersecretary for European Areas), paid a visit to the Bewysburo in May 1954 and discovered that the filing and processing of the records of tax payments and movements coming in from the regions had fallen behind by 'at least a month'. Without up-to-date records in these two areas the Bewysburo was incapable of its essential functions: calculating outstanding tax, issuing writs to tax defaulters, determining the current location of individual workers, and determining the status of individual African voters for the handful of remaining Senate representatives (which were determined by tax payments). The lag in processing, Heald believed, was prompted by the failure to fill the posts of two 'Women Supervisors' and eight assistants, frozen by the Staff Branch of the NAD in April. He demanded that the posts be reinstated, but his effort had little effect. The processing of the tax and movement records of the Bewysburo was never again up-to-date. The great avalanche of records pouring in from the districts simply overwhelmed the clerical capacity of the Bureau in Pretoria.[18]

As the number of mobile teams working in the countryside increased from eight to twelve, and the total of registered workers crept towards the one million mark in the middle of 1954, an entirely unanticipated set of archival effects began to assert themselves in the Bewysburo. The number of population registration cards being returned from the district offices of the NAD on a monthly basis exceeded the number of books being issued. By July 1954, some 80,000 of these 'C26' cards were arriving every month. The backlog of unprocessed cards duly grew to match the number of books being issued. Simply preparing the new C26 cards in anticipation of the issuing of the books required the labour of three

[17] CAD NTS 9793, 1027/400(9) Matters referred to U/S (E.A.) Central Reference Bureau. Turton to 'Cecil'. 10 March 1954.
[18] CAD NTS 9793, 1027/400(9) Matters referred to U/S (E.A.) Central Reference Bureau. Under Secretary (European Areas) to Under Secretary (Staff and Administration), 'Central Reference Bureau: Staff' c. May 1954.

clerks. And, as Turton had blithely proposed in his original plan, the Bureau had to take on the task of providing the tax payment details on the Court writs that began to arrive en masse. (In three weeks in August 1954, the Bureau had to process 17,000 of these writs.) The increased pool of registered bearers of the book resulted in one other unanticipated difficulty: applications for the issue of duplicated books to replace lost, stolen or abandoned books began to demand an increased share of the clerical capacity of the Bureau. These duplicate applications, numbering around 3,000 per month by the middle of 1954, imposed a vastly more complex processing task because they had, first, to be checked against the original application.

In addition to the fifteen mobile teams in the field, Turton now requested the staff size of the Tax and Movement section be doubled from thirty-five to sixty-seven clerks. His request said nothing about the prospects for the Finger Print section. Indeed, from the beginning of the year Turton had been attempting to prepare his supervisors for the disturbing idea that the Bureau was still viable in the absence of an adequately functioning Finger Printing section. With less than a quarter of the total male population registered, and the pace of registrations slowing in the countryside, the Bureau that Eiselen had described as 'a sound mechanism without any cost' was beginning to look berserk and expensive.[19]

On the Highveld the individual Mobile Teams had to work harder and longer to register diminishing numbers of remaining workers. Once the registration of workers in the large industrial employers of the Witwatersrand was completed, teams were sent into the country-side of the Transvaal (and then Natal) to register workers on farms, the small towns and the reserves. The average monthly registrations peaked at 80,000 in July 1954. Thereafter, even as more and more teams were despatched to the field, the rate of registrations slowed. By the end of February 1955, with fourteen teams in the field, the average number of registrations had declined to 61,000 per month.

Travelling along the battered roads in the platteland and the reserves during the summer rains, the Mobile Teams were forced to advertise their itineraries well in advance in order to ensure that workers, farmers

[19] CAD NTS 9793, 1027/400(9) Matters referred to U/S (E.A.) Central Reference Bureau. Brand, G.A. Hoofclerk, Bewysburo, to Director, Sentrale Bewysburo, 18 August 1954 and Director, Central Reference Bureau, Native Affairs Department to Heald, C.A. Undersecretary, European Areas, Department of Native Affairs. Central Reference Bureau – Staff, 31 August 1954. CAD NTS 9794, 1027/400(13) Central Reference Bureau. Forging of Reference Books. Director, Central Reference Bureau. 'Mobile Teams for the issue of Reference Book', 18 January 1954.

and officials would turn out on the days of their arrival. Often they had
to contend with magistrates and farmers who had inflated the number
of eligible workers in their districts. And many of the men who formed
part of these estimates were actually migrants who had already secured
a Dompas on the Witwatersrand. The result was that the teams often sat
idly in the countryside, and registration returns were ever diminishing.

Yet more troubling was the fact that two full years into the project,
only the Transvaal, a few districts in Natal and one or two towns in the
Free State had been covered. Just 1.3 million African men had been registered, but that seemed a strangely large proportion of the estimated
2.5 million African men that the NAD estimated lived throughout the
country. The great scientific success of the factory registrations of 1953
had begun to dissolve into the chaos of registering the thinly scattered,
mobile populations in the countryside.

### Crisis

The Bureau under its new director, Ramsay, responded to this unanticipated decline in the rate of registrations by restating an earlier request
for an expansion in the size of the staff budgeted for the field teams.
This request prompted Verwoerd's first realisation that the Dompas and
the Bewysburo were not turning out at all as he had planned. Writing to
Eiselen a month later he noted that Turton had originally advised him
that 'alle bewysboeke' would be distributed in nine months. In presenting the bill to parliament, Verwoerd had been sceptical about Turton's
claims, and he had duly doubled the amount of time required to complete the process to eighteen months. Now, he complained, 'sal meer as
drie jaar verloop het met registrasie van mans alleen [more than three
years will have passed in the registration of men alone]'.[20] The 'whole
situation' demanded reconsideration.

In the same memo Verwoerd also dragged the increasingly unworkable
fingerprinting effort from the place in the deep shadows allocated for it
by the two Bewysburo directors. Simply catching up with the classification of the outstanding fingerprints, he remarked, will take 'jare der jare'.
And it was Verwoerd who first articulated the obvious, devastating, truth.
'The fingerprint project seems totally worthless', he explained matter-of-
factly, 'even though it costs a great deal.' Urging his senior officials to see
these 'issues in the face', Verwoerd wanted answers to questions that the
others would not ask: 'When we were busy with the Bill, the Department

---

[20]  CAD NTS 1027/400 Naturelle Sake. Bewysburo (b). Verwoerd, H.F., Minister of Native
Affairs to Secretary of Native Affairs, 28 April 1955.

suggested that fingerprints were very important – what value have they had so far in practice?' Was the Bewysburo scheme a success? Was it better than the previous regime? Did it save money?[21]

Once the minister had broken the silence about the feasibility of the project, other senior officials began to tear it to pieces. The following day, in response to Verwoerd's questions, the staff branch of the NAD produced a devastating summary of the history and prospects of the Bureau. The registration of men, alone, looked set to take three years longer than the original estimate of nine months. Of the more than 1.2 mllion prints that had been collected between March 1953 and March 1955, just 300,000 had been classified. The Bewysburo had, after much bitter experience, discovered that fingerprint 'Experts' were not obtainable and 'even assistants were scarce'. In stark contrast to Francis Galton's late nineteenth-century boasts, it was taking the Bureau six months to teach novices the basics of fingerprint interpretation, and a full year before they were able to undertake anything like a meaningful load. Nor was it proving possible to keep them on the job for any length of time. As the books were distributed so the number of queries relating to duplicate books, movements and taxes increased proportionately and the failure of the fingerprinting classification and indexing effort was likely to produce further confusion.[22]

Without waiting for more information from the NAD, Verwoerd resolved the following day to push on regardless of the fate of the fingerprint classification project. 'The pace of Bewysboek distribution cannot depend on the progress with the fingerprints', he observed, 'because then it will be impossibly slow.' Drawing on the repeated assurances that Turton had offered about the viability of the Bewysburo in the absence of a working fingerprint section, Verwoerd instructed the NAD that there could be no turning back: 'the organisation must be such that the delays in the fingerprint section do not cause confusion or render the system useless'.[23]

In his response to the minister's furious questions (most of which were prompted by his being made to seem a fool before his white public), Ramsay sought to highlight the widespread appeal, and prestige, of the new books. All officials in the NAD, the municipalities and the police force acclaimed the new system as an 'unqualified success'. And

[21]  CAD NTS 9792, 1027/400 Naturelle Sake. Bewysburo (b). Verwoerd, H.F., Minister of Native Affairs to Secretary of Native Affairs, 28 April 1955.

[22]  CAD NTS 9792, 1027/400 Naturelle Sake. Bewysburo, 1955–6. L. Smuts to Verwoerd, H.F., Minister of Native Affairs, 29 April 1955.

[23]  CAD NTS 9792, 1027/400 Naturelle Sake. Bewysburo, 1955–6. Verwoerd, Minister of Native Affairs, to Eiselen, Secretary for Native Affairs, 30 April 1955.

white employers, especially the 'farming community', were ecstatic. The
Bewysburo had already established itself as the central cog in Verwoerd's
new legislative machine. It would not, Ramsay noted, have been possible to introduce the Labour Bureau system without the Bewysboekies.
Crucially, the new system allowed 'control of natives in the cities where
the books have already been distributed to move from practically nil to
a high level'. And there were the new tax revenues – the new system had
produced a half-million pound increase in each of the years of its operation in the tax payments from African men.

Ramsay was not yet convinced that the fingerprinting effort was a dead
loss. All unclassified fingerprints were indexed by the new 'persoonsnom-
mer' which, after March 1954, was issued along with the 'bewysboek'
to every African 'from registration to death'. This meant that the function of the fingerprint catalogue as the unique index of the Bewysburo
was supplanted by the new ID numbers. It also meant that the fingerprints, movement, employment and tax history of any individual could
be traced – as long as the ID number was known.[24] Of course, in the
absence of a working fingerprint classification, all that stood between an
individual and multiple 'persoonsnommers' was the Bewysboek photograph. So the Bureau could not abandon the idea that the fingerprint collection would eventually be brought up to date. Fully 130 officials were
employed on the effort within the Bewysburo by the middle of 1955. But
the work was intolerable and badly paid. Of the ninety-six people hired
for training as Fingerprint Experts in 1954, just twelve stayed on with the
Bureau. This left the Bureau with the capacity to classify a thousand sets
each day, or just 300,000 sets of fingerprints per year.

In his effort to persuade his austere minister of the continuing value of
this project, Ramsay pointed out that he had been in contact with J. Edgar
Hoover, Director of the US Federal Bureau of Investigation. Comparing
the activity of his Bureau with the FBI, he boasted to Verwoerd, 'our
daily production per person is gratifying especially as our personnel are
still being trained (31 individuals have less than 12 months service)'.
The panoptic fantasy did not die easily. Awash with huge backlogs in
Tax and Movement data, and with plans in place for an exact replica
of the Bureau to handle the processing of African women, Ramsay was
still planning to spend the next ten years catching up with the currently
registered set of African males. Needless to say, they never managed it.
By the late 1990s the collection of forty million fingerprints maintained

---

[24] CAD NTS 9793, 1027/400(9) Matters referred to U/S (E.A.) Director of Native Affairs
Central Reference Bureau, 12 January 1955.

by the Department of Home Affairs in Pretoria was estimated to contain between four and five million duplicate sets of fingerprints.[25]

## Catastrophe

In the absence of the fingerprints as the unique index of the Bewysburo everything now hinged on the integrity of the books themselves. Just as the planning around the implementation of a foolproof fingerprinting index had turned out, under the excruciating pressure of the national registration effort, to be hopelessly naive, so too did the state's ambitions for laminated documents. Towards the end of 1954, as the registration effort was being sucked into the mire on the platteland, Verwoerd began to pick up the first signs that his Bewysboekies were indeed susceptible to the age-old faking and forging practices of South African working-class life.

The first signs of renewed subversion dealt not so much with forgery, but with fakery. Just before Christmas in 1954 *The Sunday Times* ran an extravagantly worded story about a man called, appropriately enough, Government Hootman, who had been arrested in the industrial town of Vereeniging and convicted for failing to carry a pass 'even though he believed he had one'. The story turned on the significance of the imitation leather booklet carried by Hootman. It was issued by the 'Educational Centre for Demonstrations and Guidance For Natives in their trades and professions with the object to do away with idle, disorderly persons in communities'. The booklet bore all the official instructions, stamps and endorsements of the original Dompas, with some improvements. It called on all 'All Authorities and Pass Officers' to allow the bearer 'who has the centre's permission, to touch upon or pass through places named above'. And it warned ominously of 'serious penalties and payment of heavy damages against one hindering or obstructing the holder in any way'. After getting his hands on the original forgery, Eiselen was little impressed with the dangers of this particular booklet, or with this kind of forgery.[26]

The minister was less sanguine. Although the department had assured him during the drafting of the bill, he reminded Eiselen, that 'these

---

[25] CAD NTS 9792, 1027/400 Naturelle Sake. Bewysburo, 1955–6. Director, Central Reference Bureau to Secretary of Native Affairs. 21 May 1955. See Keith Breckenridge, 'The Biometric State: The Promise and Peril of Digital Government in the New South Africa', *Journal of Southern African Studies*, 31, no. 2 (2005): 267–82.

[26] CAD NTS 9794, 1027/400(13) Central Reference Bureau. Forging of Reference Books. Eiselen, W.W.M. to Verwoerd, H.F. 'Bogus Pass Racket: Thousand Duped', 3 February 1955.

books, in contrast with the previous passes, could not be falsified', he was little mollified. It was evident 'that forgeries are taking place that are only occasionally detected' and he called for 'very strong measures to be provided for forgers' in the upcoming amendments to the original bill. And, mindful of the increasingly unworkable fingerprinting system, Verwoerd wanted the police to see the 'seriousness of forgery', to hunt down the source of the documents with 'great seriousness' and to apply, yet again, 'very strong measures'.[27]

Just two weeks later the first real signs of the sort of forgery that Verwoerd suspected came to light in a Johannesburg magistrate's court. An enterprising forger had removed the lamination, identity card and photograph from a stack of Bewysbooks. He had successfully and completely erased all the text on the paper inside the laminated identity cards, and inserted new personal details, photographs and cellophane. And he was about to sell the documents to a new set of bearers when the police arrested him. These events prompted the Bewysburo to re-examine its untouchable documents. One of the members of the Fingerprint section, presumably to distract himself from the tedium of his work, volunteered to dismantle a Bewysboek. He managed to do so – notwithstanding the South African Bureau of Standards' scientific assurance that it was impossible – simply by steaming it.

Shortly thereafter the NAD ended its experiment with laminated documents. The personal identity card (for whites as well as blacks) was later abandoned, and replaced by a new Dompas, which integrated the personal details and photograph of the bearer onto the cover of the book. The law was amended in 1955 to make provision for Verwoerd's 'very strong measures for forgers' and magistrates and the police took a much sterner view of the mangling of the Dompas.

The new books simplified some of the document processing, but they did little to solve two outstanding difficulties. The first of these was the very widespread defacement of the Bewysbooks. Pages from books that had been endorsed with permission to enter or work in one of the cities were frequently removed and pasted into books without the necessary endorsements. Alternatively, pages that were intended to force workers to remain either on the farms or in the labour reserves were often removed and replaced with blank pages from another book. Towards the end of 1955, the Bewysburo began to resort to the kind of bureaucratic

---

27 CAD NTS 9794, 1027/400(13) Central Reference Bureau. Forging of Reference Books. Verwoerd, Minister of Native Affairs to Eiselen, W.W.M. 'I/s: Koerantberig: Bogus Pass Racket: Thousand Duped.' 20 December 1954. Verwoerd's favourite phrase here is 'baie streng strawwe'.

inscription and earmarking that have been used to protect official documents throughout the documentary epoch: punching holes through the middle of the Bewysbook and writing the identity number over any endorsements.[28] But these retaliatory forms of embellishment could do little to prevent the fundamental source of the weakness of the Dompas – the duplication of books. Without the unique fingerprinting index, there was almost nothing that the Bewysburo could do to prevent people from discarding unwanted books, or from applying for new books under multiple identities. By the end of 1954, with just over half-a-million books issued, the Bureau had already processed 29,000 requests for duplicate books – a proportion of the total that was massively higher than anything anyone had anticipated. The director of the Bureau admitted in October 1955 that 'some natives have received as many as nine reference books'. The first list of suspicious Bewysbook numbers that had been sent to the police in March 1954 had contained 2,300 names. Thereafter, as the Commissioner of Police explained in 1955, the list of numbers grew so rapidly that it was impossible even to review them.[29]

By December 1955, the Bewysburo was in a state that can fairly be described as chaos. Processing of all of the major databases lagged so far behind that they could not provide the basic functions required of them. The backlog of tax and movement data, for example, was nine months in arrears, which meant that the NAD was unable accurately to determine the voters' roll for the nominal Senate representatives still retained for Africans who had completed their tax payments. The silicosis records, which provided a statutory function allocating benefits to mine-workers afflicted by deadly and disabling lung diseases, were similarly out of date. At this point the Director of the Bureau estimated that it would take a decade, and a substantial expansion of the existing staff, to bring the fingerprinting effort under control. Indeed, if the collection continued to grow 'at the present rate another building the size of Heyvries building will be required to house finger prints alone'.[30]

[28] CAD NTS 9793, 1027/400(9) Central Reference Bureau. U/S (E.A)'s papers regarding mechanisation. Young, Y. Director, Central Native Reference Bureau to Undersecretary (European Areas). 6 October 1955.
[29] CAD NTS 9793, 1027/400(9) Matters referred to U/S (E.A.) Central Reference Bureau. Director of Native Affairs Central Reference Bureau, 12 January 1955. CAD NTS 9793, 1027/400(9) Central Reference Bureau. U/S (E.A)'s papers regarding mechanisation. Young, Y. Director, Central Native Reference Bureau to Undersecretary (European Areas). 6 October 1955. CAD SAP 494, 15/2/52 Uitreiking van Bewysboekies aan Bantoes, 1953–7, 'Bewysboeke-Duplikate (Memorandum)', 23 June 1955.
[30] Ibid.

As the decade moved to a close, key individuals within the Department of Native Affairs, the Bureau itself, and private firms began to put forward the case that 'mechanisation' was the only possible remedy. They meant by this that powered mechanical filing cabinets, microfilming and the widespread introduction of punch-card readers and tabulators could solve the backlog of tax, movement and silicosis records. There was never any intention to 'mechanise' the classification and cataloguing of fingerprinting records – that process only began in earnest in the 1980s. Mechanisation began at precisely the moment that the Bewysboek began to face another much more formidable enemy – refusal and burning at the hands of women who were being forced to bear passes for the first time, and a militant national campaign of resistance to the new Pass Laws organised by the Pan Africanist Congress.[31] Both of these stories have been told – in some detail – elswhere and their combined effects did little to change the fact that the Dompas system was fundamentally broken.

The South African Police issued repeated instructions in the late 1950s and early 1960s, reminding its members not to refer requests for missing African suspects to the Bewysburo. 'In the beginning the Bewysburo was helpful in this area', Captain Jooste of the SAP's Criminal Bureau explained to his boss in 1959, 'but they were so overwhelmed with queries that they could simply not process them.' The result was that the Bureau could not even assist the police with the 'tracking of deserters'. Far from providing the all-embracing panopticon that Turton and many imagined of the Dompas system, the Central Reference Bureau could not even initiate the most basic functions of the old documentary pass system. Yet, while it may not have had the disciplinary effects that were intended for it, the Bewysburostelsel certainly had devastating effects on the majority of South Africans.[32]

As the panoptic functions of the Bewysburo disintegrated at the beginning of the 1960s, with an increasingly capricious documentary regime and widespread hostility and resistance to the bearing of passes, so the NAD began to call for increasingly rigorous policing. The Johannesburg Bantu Affairs Commissioner complained bitterly in September 1960 that the reluctance of the police to 'ask for the Bewysboek [was ruining] the work of the last three years. The Bantu are streaming to the cities.'

[31] Julia C. Wells, *We Now Demand! The History of Women's Resistance to Pass Laws in South Africa* (Johannesburg: Witwatersrand University Press, 1993), 110–26; Tom Lodge, *Black Politics in South Africa since 1945* (Johannesburg: Ravan Press, 1983).

[32] CAD SAP 494, 15/2/52 Uitreiking van Bewysboekies aan Bantoes, 1957–61. Adjunk-Kommissaris, Afdeling Transvaal, SA Polisie. 19 September 1959 and Commissioner, South African Police, Pretoria. 'Force Orders (General)', 27 May 1962.

The Department, with Prime Minister Verwoerd's support, called for much more vigorous police intervention, and for a 'sharpening' of influx control. By 1962 the Dompas had duly become the key instrument of a brutally enforced white supremacy – a mechanism of capricious policing, mass arrest and imprisonment. The average of monthly arrests under the Bewysbook regime (a new category called 'Bantu Control Measures' by the SAP) rose to 49,000 in the middle of 1962. There is a macabre symmetry in the figures for monthly registrations in the 1950s and monthly arrests in the 1960s.[33]

## Consequences

In one respect at least the Bewysburo did exactly what it was intended to do: it tilted the already skewed labour relationship in South Africa massively in favour of employers. The requirement that workers on farms secure their employer's written consent to cancel the effects of the 'Labour Tenant' stamp bound them to these jobs in the countryside much more effectively than any of the previous paper passes. The permanent Dompas had a similar effect in the suburban homes of whites. In the late 1950s African women were forced into state regulated employment, usually as servants. The Bewysburostelsel 'drove large numbers of women into the wage labour market' as domestic workers because an employer's endorsement in the Dompas was one of the only ways to retain access to the cities. 'For the first time', Julie Wells observed, 'the Johannesburg City Council in 1959 reported no shortage of labour for domestic service.'[34]

In the countryside the Bewysburostelsel worked in concert with ongoing 'black spot' removals to swell the population bottled into the reserves far above the subsistence threshold. Greatly increased population densities knocked away the last supports of small-scale farming for most rural households.[35] So it was that when Verwoerd suddenly turned in May 1959 to a Bantustan strategy that would confer nominal constitutional independence on the African 'homelands' he was attempting the impossible. In the place of the well-ordered programme of rural development proposed by Nationalist intellectuals in the Tomlinson commission Verwoerd embarked on a programme of rural ghettoisation, achievable

[33] CAD SAP 494, 15/2/52 Uitreiking van Bewysboekies aan Bantoes, 1957–61, Waarnemende Kommissaris, Suid Afrikaanse Polisie na Alle Afdelingskommissarisse, SA Polisie, Republiek van Suid-Afrika en Suidwes-Afrika, 1962/07/10.

[34] Wells, *We Now Demand!*, 125.

[35] Peter Delius, *A Lion Amongst the Cattle: Reconstruction and Resistance in the Northern Transvaal* (Oxford: James Currey, 1996), 142–5.

only through violence and the systematic corruption of local authorities. It was, famously, the Dompas that allowed him to do this. It provided him with a blunt instrument to lock the majority of African men and women into the deteriorating subsistence economies of the patchwork of so-called Independent and Self-governing states.[36]

Yet there was little about the ideas associated with Apartheid, or Afrikaner nationalism, in the 1940s and 1950s to prepare people for the policy of radical constitutional separation that Verwoerd imposed on the country in 1959. The National Party stunned itself, and the country, in 1948, taking power with a five-seat majority and just 40 per cent of the white vote. At that time the key intellectual advocates of Apartheid, like Werner Eiselen, understood it as a programme of systematic racial separation. He argued for 'the separating of the heterogeneous groups ... into separate socio-economic units, inhabiting separate parts of the country, each enjoying in its own area full citizenship rights'.[37] His idea of Apartheid was heavily influenced by the moral and cultural importance of linguistic community, but there was no suggestion of territorial separation.[38] And within the broad ranks of the National Party Eiselen was an isolated, idealistic, scholar.

Afrikaners in general in the 1950s, as Giliomee has shown, were preoccupied with the cultural and demographic achievements of a resurgent nationalism, and they had little concern for large-scale racial engineering.[39] This lack of interest in the project of territorial separation was clearly evident in Verwoerd's rejection of the Tomlinson commission of inquiry into the 'Socioeconomic development of the Bantu Areas', which gathered evidence for five years on the problem of making the labour reserves self-sustainable. The five-volume report proposed the division of the country into 'seven historiological nuclei or heartlands, based on the linguistic and territorial divisions of the country', and an extravagant programme of spending to bolster the homeland economies of these regions. It was Verwoerd who dismissed both the legal and the fiscal proposals because they threatened to temper white prosperity in general, and Afrikaner nationalist accumulation in particular.[40] He received the commission's report a year into his tenure as Minister of Native Affairs, when the Bewysburo system still held out the promise of complete surveillance of the identity and movements of all Africans.

[36] Deborah Posel, *The Making of Apartheid, 1948–1961: Conflict and Compromise* (Oxford and New York: Clarendon Press and Oxford University Press, 1991), 226–55.
[37] Cited in Hermann Giliomee, *The Afrikaners: Biography of a People* (Cape Town: Tafelberg, 2003), 500.
[38] Evans, *Bureaucracy and Race*, 228.
[39] Giliomee, *The Afrikaners*, 491.    [40] *Ibid.*, 517.

Things had changed by the end of the decade. In May 1959 Verwoerd, in his new post as prime minister, introduced the Bantu Self-Government Bill, setting in motion the territorial separation of ten 'Bantu Homelands'. He brought the bill to parliament without consulting his own party caucus, and to the evident astonishment of his old departmental secretary, Werner Eiselen, who weeks before had published an article pronouncing that the political authorities in the African reserves would never be granted independence.[41] It was this stunning reversal of his own policy that marked the beginning of the demented social engineering of Grand Apartheid.

A number of explanations have been offered for Verwoerd's change-of-mind. The first set of these hinges around the political climate of decolonisation, graphically represented inside South Africa by Harold Macmillan's 'Winds of Change' speech to the South African parliament in February 1960. Recently, Hermann Giliomee has suggested that Verwoerd was encouraged to develop a more credible Bantustan strategy by a series of meetings with the UN Secretary-General, Dag Hammerskjöld, in January 1961.[42] There is certainly no doubt that, in the years that followed, Verwoerd sought to adorn Grand Apartheid with this 'anti-colonial' mantle. 'In the light of the new spirit and the pressures exerted and the forces which arose after the Second World War it is clear', he told parliament a year before his assassination, 'that no country could continue as it did in past years.'[43]

Another key line of explanation hinges on the explosion of political protest that surrounded the massacre of Dompas protestors at the Sharpeville police station in March 1960. The argument here is that the pressure of African political protest grew steadily from the start of the 1950s, culminating in the protests against the introduction of passes for women in 1958, violent protests in Cato Manor near Durban in 1959, and the PAC's national campaign against the pass laws in the early 1960s. Nationwide protests coincided with the first assassination attempt on Verwoerd on the day after the official banning of the ANC and the PAC, which would see him incapacitated in hospital while the country reeled from the local and international effects of the Sharpeville massacre. This crisis prompted a dramatic reordering of the lines of political conflict within Afrikaner nationalism, and white politics more generally. When

---

[41] Dan O'Meara, *Forty Lost Years: The Apartheid State and the Politics of the National Party, 1948–1994* (Johannesburg: Ravan Press, 1996), 73; Evans, *Bureaucracy and Race*, 232. and W.W.M. Eiselen, 'Harmonious Multi-Community Development', *Optima* 9, no. 1 (1959).

[42] Giliomee, *The Afrikaners*, 531.

[43] O'Meara, *Forty Lost Years*, 74.

Verwoerd returned from hospital he insisted on a policy of no-compromise. Within months, his unyielding position had succeeded in breaking the back of black protest, and restored much of South Africa's international economic standing.[44]

There can be little doubt that all of these events shaped the form of the state in South Africa after 1960, but they cannot account for Verwoerd's specific decision to turn to the Bantustan policy as his strategy for the maintenance of white supremacy. For Verwoerd's introduction of the Bantu Self-Government Bill pre-dated all of these events by at least a full year. In examining the turn to the Bantustan policy, it is useful to keep in mind the key elements of Verwoerd's political temperament. He was a politician for whom 'obstacles in human nature, must give way to regulation and systematization', the leading newspaper editor of his time observed, 'The ideal must be imposed on the society'.[45]

It seems reasonable to suggest that Verwoerd made the decision to dismember the state after taking the prime minister's office in 1958 because his own earlier effort to build a panoptic administration around the Bewysburo – an effort to sweep away the documentary chaos of the segregationist era and establish an efficient, centralised hold on the vital details of the black population – lay in ruins. Verwoerd, the most determined centralist since Lord Alfred Milner, turned to the Bantustan strategy as a conscious effort to break the administrative unity of the South African state. In so doing, he sought to subcontract problems of control to weakly regulated African subordinates in fictionally independent states, whilst paying little regard to their actual administrative performance. It is the legacy of this dismantling that underpins many of the most intractable problems of state capacity in contemporary South Africa.

## Conclusions

Most of the contemporary scholarship on Apartheid, especially on the period of Hendrik Verwoerd's government, has stressed the state's great success in regulating the lives of its subjects. 'The system of Apartheid', Cherryl Walker observed some twenty years ago, 'can be summed up in one word – control.'[46] Influx control, the fingerprinting effort and the Bewysburo are at the heart of this perception of Apartheid as the

---

[44] Posel, *The Making of Apartheid, 1948–1961*, 245–6; Giliomee, *The Afrikaners*, 522–5.

[45] Cited in Giliomee, *The Afrikaners*, 520.

[46] C. Walker, *Women and Resistance in South Africa* (New York: Monthly Review Press, 1991), 123.

informated police state.[47] Ivan Evans suggested that the labour bureaux systems equipped the police with sophisticated surveillance on the activities of the state's abundant enemies.[48] And there is an element of truth here, at least in so far as the state affected the lives of individuals. But most of the historians and political scientists who have written about this period have taken the state at its own public word, and massively overstated its coercive and surveillance abilities in the process.

The Bewysburo project – and the brutal policing required to enforce it – associated the Apartheid project with an iconography of racist abjection that provided its opponents around the world with very powerful and moving political testimony. These images emerged from many sources, beginning with Margaret Bourke-White's *Time-Life* images of South Africa in 1949, they were sustained by *Drum* photographers like Peter Magubane who captured both the violence of policing and the realities of segregated urban life; by the 1980s many photographers – notably Santu Mofokeng, Guy Tillim and Paul Weinberg – were producing images of the state's violence that echoed Fanon's descriptions of colonialism as a Manichean world of intolerable violence.[49] Perhaps more than the work of any other single photographer, it was Ernest Cole's *House of Bondage* that suggested the totalitarian quality of the Bewysburo regime (see Figures 5.1, 5.2 and 5.3).[50] It was these images that fostered the Anti-Apartheid Movement around the world, and they have – not least because of their prominence in the Apartheid Museum in Johannesburg – encouraged an understanding of the state as a successful surveillance order.[51] Much of that sense of control is misplaced.

Yet the Bewysburo also placed South Africa on a path of biometric government that it could not escape. For decades afterwards the state remained doggedly committed to the effort to capture and classify the

---

[47] Mark Sher, 'From Dompas to Disc: The Legal Control of Migrant Labour', *Crime and Power in South Africa*, ed. Dennis Davis and Mana Slabbert (Cape Town: David Philip, 1985), 79.

[48] Evans, *Bureaucracy and Race*, 90.

[49] Holland Cotter, '"Rise and Fall of Apartheid" at Center of Photography', *New York Times*, 20 September 2012, www.nytimes.com/2012/09/21/arts/design/rise-and-fall-of-apartheid-at-center-of-photography.html; Frantz Fanon, *The Wretched of the Earth: A Negro Psychoanalyst's Study of the Problems of Racism and Colonialism* (New York: Grove Press, 1966).

[50] Ernest Cole, *House of Bondage By Ernest Cole (A South African Black Man Exposes in His Own Pictures and Words the Bitter Life of His Homeland Today)*, ed. Thomas Flaherty (New York: Random House, 1967); Celia W. Dugger, 'Ernest Cole's Photographs Exhibited in South Africa', *New York Times*, 17 November 2010, www.nytimes.com/2010/11/18/arts/design/18cole.html.

[51] See also NARMIC, *Automating Apartheid: US Computer Exports to South Africa and the Arms Embargo* (Philadelphia: NARMIC, 1982).

5.1 Photograph of a worker being fingerprinted, published in Ernest Cole's book *House of Bondage*

5.2 Photograph of a queue outside a Labour Bureau, published in Ernest Cole's book *House of Bondage*

5.3 Photograph of a youth being stopped for a pass, published in Ernest Cole's book *House of Bondage*

fingerprints of its African subjects – notwithstanding the ongoing impossibility of the project.[52] By the early 1980s – as we will see next – it had resolved to expand this project of fingerprint registration to whites, Indians and Coloureds. That decision coincided with the development of new technologies of automation that – slowly and fitfully – changed the horizons of possibility and the viability of the biometric state.

[52] Paul N. Edwards and Gabrielle Hecht, 'History and the Technopolitics of Identity: The Case of Apartheid South Africa', *Journal of Southern African Studies* 36, no. 3 (September 2010): 619–39.

# 6  Galtonian reversal: apartheid and the making of biometric citizenship

In the previous chapters we have seen that Francis Galton's plans for a biometrically ordered society were distinctively elaborated in South Africa in the century after his death, culminating in the massive project to build the Bewysburo in the 1950s. Very recently Galton's blueprint has escaped from the special conditions in South Africa, finding its way in to almost every one of the former colonies of the European empires, with centralised ten-print biometric identity registration filling the administrative void left by the weak states and limited ambitions of colonial governments. Many researchers have pointed out that a new biometric infrastructure of citizenship has begun to emerge in the territories of the former European empires.[1] The mechanisms of these new technologies of citizenship are partly financial – involving a variable suite of conditional cash transfers for the very poorest individuals – and partly administrative – requiring identifying fingerprints from recipients in the absence of the documentary forms of identification that have historically supported civil registration. In Mexico, Brazil, India and dozens of other countries, the state's new interest in cash grants for individuals is fostering the development of very ambitious, centralised, biometric population

---

[1] J. Hanlon, 'It Is Possible to Just Give Money to the Poor', in *Catalysing Development? A Debate on Aid* (Oxford: Blackwell, 2004), 181; Joseph Hanlon, 'Just Give Money to the Poor', Conference Item, 22 April 2009, http://oro.open.ac.uk/20887/; J. Hanlon, D. Hulme and A. Barrientos, *Just Give Money to the Poor: The Development Revolution from the Global South* (Sterling: Kumarian Press, 2010); M. Garcia and C. Moore, *The Cash Dividend: The Rise of Cash Transfer Programs in Sub-Saharan Africa* (Washington, DC: The World Bank, 2012), www-wds.worldbank.org/external/default/WDSContentServer/WDSP/IB/2012/03/01/000386194_20120301002842/Rendered/PDF/672080PUB0E PI0020Box367844B09953137.pdf; A. Gelb and C. Decker, 'Cash at Your Fingertips: Biometric Technology for Transfers in Developing Countries', *Review of Policy Research* 29, no. 1 (2012): 91–117; Alan Gelb and Julia Clark, 'Identification for Development: The Biometrics Revolution' (Centre for Global Development, 28 January 2013), www.cgdev.org/content/publications/detail/1426862/; James Ferguson, 'What Comes After the Social? Historicizing the Future of Social Assistance and Identity Registration in Africa', in *Registration and Recognition: Documenting the Person in World History*, ed. Keith Breckenridge and Simon Szreter, Proceedings of the British Academy 182 (Oxford University Press, 2012), 495–516.

registers which will amount to a radical break with long-established architectures of state administration in each of these countries.[2]

It is intriguing how much of this new infrastructure of biometric government was suggested by Francis Galton over a century ago. He first proposed in 1891 that fingerprinting would be an 'invaluable adjunct to a severe passport system'.[3] He was also the first to recommend, in a letter to *The Times* in 1893, that fingerprints would be invaluable in identifying bank account holders.[4] And building an infrastructure of biometric registration in the territories of the empire, as we have seen, was always the heart of his project. Fingerprint registration was not necessary, as he put it, 'in civilized lands for honest citizens' where identification was regulated by signatures, photographs and personal introductions. But, with the imperial origins of the Tichborne scandal – where a claimant to a vacant baronetcy emerged after a decade in Australia – clearly in mind, Galton thought that it was necessary 'when honest persons travel to distant countries' to create a register of emigration.[5] Many years later Galton's grim and brilliant disciple, Karl Pearson, lamented the failure of the centralised biometric register of identity in Britain, but he was aware that the idea was doomed by the 'taint of criminality'. In fact the special reputational problem that fingerprinting confronted in Britain – in addition to the degrading marking of the body and the association with policing – was that it bore the taint of imperial coercion, in its German and South African flavours.[6]

---

[2] Nelson Arteaga, 'Biological and Political Identity: The Identification System in Mexico', *Current Sociology* 59, no. 6 (1 November 2011): 754–70; David Murakami Wood and Rodrigo Firmino, 'Empowerment or Repression? Opening up Questions of Identification and Surveillance in Brazil through a Case of "Identity Fraud"', *Identity in the Information Society* 2, no. 3 (1 December 2009): 297–317; The Fletcher School, Tufts University and Bankable Frontier Associates, 'Is Grandma Ready for This? Mexico Kills Cash-Based Pensions and Welfare by 2012' (presented at the Killing Cash – Pros and Cons of Mobile Money for the World's Poor, The Fletcher School, Tufts University, 2011); F. Zelazny, 'The Evolution of India's UID Program' (2012), http://cgdev.org/files/1426371_file_Zelazny_India_Case_Study_FINAL.pdf; Gelb and Clark, 'Identification for Development'.

[3] Francis Galton, 'Identification by Finger Tips', *Nineteenth Century* 30 (August 1891): 303.

[4] Cited in Karl Pearson, *The Life, Letters and Labours of Francis Galton: Correlation, Personal Identification and Eugenics*, vol. 3A (Cambridge University Press, 1930), 158.

[5] Francis Galton, *Finger Prints* (London and New York: Macmillan and Co., 1892), 148; Galton, 'Identification by Finger Tips', 10; Edward Higgs, *Identifying the English: A History of Personal Identification, 1500 to the Present* (London and New York: Continuum, 2011), 146.

[6] Pearson, *The Life, Letters and Labours of Francis Galton*, vol. 3A, 159; Jon Agar, 'Modern Horrors: British Identity and Identity Cards', in *Documenting Individual Identity: The Development of State Practices in the Modern World*, ed. Jane Caplan and John Torpey (Princeton University Press, 2001), 101–20; Edward Higgs, 'Fingerprints and Citizenship:

Galton argued, from the outset, that biometric registration was well suited to the special features of government in the colonies where 'natives are too illiterate for the common use of signatures'; fingerprints, he argued, could be used to regulate government examinations and international movement, but they could also give the colonial state control of the payment of salaries, pensions and loans.[7] It is precisely in these latter areas that computerised biometric systems are being very widely adopted in the former colonial world.[8] The story of how this happened is complex, as we have seen: biometric registration has long been both a theoretical and a practical alternative to documentary registration in the empire, acting as a constraint on the development of forms of literary registration, and equipping the colonial state with effective and limited tools of surveillance. On the African continent the South African state served as a laboratory and model for these systems. That story fits with much of what we know about the colonial state in Africa. My object in this final chapter is to follow a more improbable process, to show how the South African system of biometric government came to deliver a special kind of welfare state that has, in turn, come to provide a model for a new global infrastructure of registration, and a new global system of biometric citizenship.

It is important and, I think, interesting to notice that – although technically derived from his specific suggestions – in almost every respect the new biometric systems are the political antithesis of Galton's eugenicism. In the first instance these new systems direct resources at the poorest populations, arguing (against Galton) that broader processes of development in fact hinge on improving the prosperity of those at the bottom of the social order. Less obviously, these new forms of biometric registration offer intensely individualised identification in place of race and caste – the brutal and cheap forms of group identification beloved by both Galton and the weak institutions of the colonial state. As things currently stand the global infrastructure of biometric cash transfers delivers very finely-grained biographical registration and recording of the financial transactions of billions of people on the planet, often closely linked

The British State and the Identification of Pensioners in the Interwar Period', *History Workshop Journal* 69 (2010): 63.
[7] Francis Galton, 'Identification Offices in India and Egypt', *Nineteenth Century* 48 (July 1900): 119.
[8] Wood and Firmino, 'Empowerment or Repression?'; Keith Breckenridge, 'The Biometric State: The Promise and Peril of Digital Government in the New South Africa', *Journal of Southern African Studies* 31, no. 2 (2005): 267–82; Keith Breckenridge, 'The World's First Biometric Money: Ghana's E-Zwich and the Contemporary Influence of South African Biometrics', *Africa: The Journal of the International African Institute* 80, no. 4 (2010): 642–62.

to powerful computerised systems for assessing creditworthiness. In this they reflect the administrative individualisation and privatisation that many scholars argue are essential to the neoliberal epoch.[9] If the current trajectory continues the new ordering that will emerge from this infrastructure will be determined by financial institutions, separating the creditworthy wheat from the financially delinquent chaff. Yet the possibility exists that biometric citizenship might do something very different, providing a language and an effective remedy for the most serious problems of inequality and injustice, directing life-saving resources to the very young, the weak and the old. In the process, these new forms of biometric payment can bypass the moralising agenda that has bedevilled welfare for centuries.[10] This change in the arrangements for the distribution of welfare outside of the wealthiest countries, as Jim Ferguson and Robin Blackburn and many others have noted, is a moment of the deepest historical significance.

How then did the forms of coercive biometric registration, organised around the control of migrant labour and the brutal regulation of African urbanisation, come to serve as the basis for a global system of poverty alleviation? This is a long and interesting story, with contributing developments in many countries, but the pivotal events took place in the Bantustan of KwaZulu in the early 1990s. In the last years of the Apartheid state, the distribution of old-age pensions for black people – especially for women – became a matter of intense significance for the Nationalist Party (desperately looking for political allies), for Mangosutho Buthelezi's Inkatha Freedom Party, and for a dispersed alliance of feminist women mobilised by a protest movement called the Black Sash. It was an odd and unstable alliance between these three interest groups – and the peculiar institutional momentum of fingerprinting identification in South Africa – that pushed one of the largest banks into developing a system of biometric cash transfers.

The new global infrastructure of citizenship is derived on the one hand from the system of biometrically delivered social grants that was invented and elaborated in the South African Bantustans. But it also relies – in India, Mexico, Brazil and dozens of other countries – on the development of a single, centralised biometric population register which, in all

[9] David Theo Goldberg, *The Threat of Race: Reflections on Racial Neoliberalism* (Malden: John Wiley & Sons, 2009), 51, 332–5.
[10] This is what Fields and Fields call for in *Racecraft: The Soul of Inequality in American Life* (London: Verso, 2012), 282–4; Anna Marie Smith, *Welfare Reform and Sexual Regulation* (Cambridge University Press, 2007) examines the very onerous forms of moral supervision in contemporary US welfare systems; on the possibility of removing moralisation from welfare, see Ferguson, 'What Comes After the Social?'

of these countries, represents an abrupt break from older and entrenched institutions of civil registration. In developing this kind of population register these societies are adopting an institution that lay at the heart of the peculiar form of the twentieth-century South African state, although admittedly for very different ends than those intended by the architects of Apartheid.

## The Population Registration Act

Population registration was the legislative cornerstone of Apartheid, the basis of the Group Areas Act, and, as we have seen in the preceding chapter, of the pass system. It was also, as Posel has shown, a powerful administrative loom in the fashioning of the enduring South African obsession with racial reasoning.[11] The proposal for a central register of racial identification long pre-dated Apartheid, but like many of the other plans for social engineering in the first half of the century, it lay unrealised in the face of administrative indecision and parsimony.[12] That changed in 1948. The Population Registration Act of 1950, which preceded both Group Areas and the Bewysburo, required every South African to secure an identity number, to register their current address and secure a racial classification. On the latter everything turned under Apartheid, yet the actual procedures of classification were carefully ad hoc, with untrained white census workers making life-altering decisions about individuals on racial criteria that were, at once, arbitrary and contradictory.[13]

In place of the unstable biological or cultural criteria that informed racial boundaries, the state built four separate population registers that fixed racial identities in perpetuity. For whites, the answers they gave to the enumerators in the 1951 census were captured on punched cards, retained by the Census Bureau and used to confirm the racial status that they claimed on the new identity cards that were issued after 1953.[14] For Coloureds and Indians the process was much more contested: data on the census forms was usually incomplete or officially suspect.

---

11 Deborah Posel, 'What's in a Name? Racial Categorisations under Apartheid and Their Afterlife', *Transformation* (2001): 50–74; Deborah Posel, 'Race as Common Sense: Racial Classification in Twentieth-Century South Africa', *African Studies Review* 44, no. 2 (2001): 87–113.
12 Posel, 'Race as Common Sense', 98.
13 Geoffrey Bowker and Susan Leigh Starr, *Sorting Things Out: Classification and Its Consequences* (Cambridge, MA: MIT Press, 1999); Posel, 'Race as Common Sense'.
14 Bureau of Census and Statistics, South Africa, *Report of the Director, 1951* (Pretoria, 1951); Bureau of Census and Statistics, South Africa, *Report of the Director, 1952* (Pretoria, 1952).

Following a tradition that dated back a century, the Apartheid state utterly mistrusted the biographical information provided by Indians, forcing them to apply in person for their identity cards using an official photographer. At these interviews Census officials were careful to insist that the applicants (most of whom were descended from indentured labourers who had arrived a century earlier) provide an original nationality.[15] The registration of Coloured people was even more contested, with only a trickle of voluntary applications for identity cards. From 1956 municipalities, mostly in the Transvaal, began to coerce Coloured people to present themselves for racial classification. This was the high moment of the humiliating bureaucratic ordeal of racial classification, with tens of thousands of individuals and families being forced onto terrifying life trajectories at the whim of Census officials.

By the end of the 1950s regional offices in each of the provinces had been established to issue identity cards and undertake 'the classification of persons of doubtful race'. While the Census Bureau concentrated on policing the confused boundaries of the creolised cities, inventing firm racial identities for whites and Coloureds in the population register, in 1952 Africans were allocated en masse to the special fingerprint requirements of the Bewysburo.

During the 1950s the population register was built on the cheap. Most of the work was undertaken in five military huts that had been attached to the main offices of the Bureau in Pretoria. They suffered from severe shortages of staff, very high rates of staff turnover and broken and worn out tabulating machines. By the end of the decade most of these problems had been addressed, with new buildings, additional staff and new tabulating machines. But, in that year, the population register was transferred out of the poorly resourced Census Bureau and into the Department of the Interior. From there, close to the centre of state power, the project of racial registration proceeded energetically.

By 1967, a generation after the passing of the initial act, the great majority of adults had been captured in one of the population registers. For the three million successful whites this meant that the temporary enumerators of the 1951 census had not disputed their self-selected race on the census forms. The only further test of race for those commonly identified as white was the submission of photographs for the laminated identity cards that were issued after 1956. For many of the one million people who were registered as Coloured, racial classification involved a much more humiliating and frightening official examination, where

---

[15] Bureau of Census and Statistics, South Africa, *Report of the Director, 1957* (Pretoria, 1957).

the legal criteria of racial acceptance and appearance were decided by officials of the Department of the Interior. And for the ten million Africans registered by the combined efforts of the mobile teams from the Bewysburo and the coercive requirements of influx control the process was both banal and horribly degrading.[16]

It was at this point, as the first cohort of sixteen-year-olds who had not been classified during the 1951 census began to apply for identity cards, that the state moved to embed a new form of racial descent in to the operations of the population register. In the determination of race before 1967 the law had been carefully crafted to avoid racial geneal-ogy (which had been used, for example, in segregationist Natal, Nazi Germany and the United States).[17] Instead, for the first decade the Act relied on pragmatic, and arbitrary, tests of communal acceptance and appearance.[18] Initially the leaders of the National Party were anxious not to make the test of whiteness a matter of descent, because they were well aware as, H.J. 'Bronkie' Bronkhorst reminded them in parliament, that few 'amongst us can beat upon his breast and proclaim he is of pure white descent'.[19] But by 1967 the logic of descent had been changed by the work of the population register. As the minister explained, those who were applying for identity cards after that date would have to answer a simple, chilling question: 'how were your parents classified?'

For whites and Coloureds, the officials built up most of the population register by carefully extracting racial selections from the responses to the original 1951 Census onto punched cards. This use of the Census data was illegal, a violation of the prohibition in the Census Act on the use of response data in 'any legal proceedings' and an obvious breach of the promise on the forms that 'no Government Department or any private person' would have access to the responses. But the Census data used in compiling the population register was quickly obscured by the applica-tions for identity cards that required a similar act of self-classification; before 1967, where the race claimed on the identity card conflicted with either the respondent or the enumerators' assessment from the 1951 Census, officials were licensed to apply the practical appearance and acceptance tests. After that date racial classification became a matter of bureaucratically determined descent, and some of this was completely invisible to the individuals being classified. The population register grew

---

[16] 'Population Registration Amendment Bill. Second Reading' (*Hansard*, 17 March 1967).
[17] Jane Caplan, 'Registering the Volksgemeinschaft. Civil Status in Nazi Germany 1933–1939', in *A Nazi 'Volksgemeinschaft'? German Society in the Third Reich* (German Historical Institute London and Oxford University Press, 2012).
[18] Posel, 'Race as Common Sense'; Bowker and Starr, *Sorting Things Out*.
[19] 'Population Registration Amendment Bill. Second Reading'.

steadily from the early 1950s, expanding by about 100,000 individuals annually from birth registration events gathered – mostly from hospitals – for whites, Coloureds and Indians, silently allocating all individuals to a racial group. 'Every person is classified at birth', the minister explained, 'but they only become aware of their classification when they apply for an identity card after their 16th birthday'. By the end of the 1960s the population register had done the impossible job, as the government constantly boasted, of allocating all South Africans to one of Verwoerd's four basic racial categories.

## The Book of Life

High Apartheid took its pure form in the years between the assassination of Verwoerd in 1966 and the international oil and domestic labour crises in 1973. Foreign investment, most of it from the United States, surged back in to the country after fleeing in the wake of the Sharpeville crisis. With the wages of migrant labourers locked at historically low levels by the combined operations of the Dompas and massive subcontinental migration to the mines, the rates of return on investment in South Africa were amongst the highest in the world. By 1970 the long boom had lifted white prosperity dramatically in real terms and relative to black people. And Afrikaner capital, for the first time, had secured its place at the high table of the mining economy.[20]

It was in this comparatively short window of self-confidence and affluence – which stands in contrast to the much longer periods of parsimony and hesitation before and chaotic reform afterwards – that the grandest technopolitical projects of the Apartheid state were born.[21] It was in 1970, for example, that Prime Minister Vorster announced the project for the domestic enrichment of uranium leading to weapons capacity.[22] The plans for the massive dams of the Orange River Development Project, which sought to re-arrange the climate of the subcontinent, were first drafted by the great imperial dam builder William Willcocks in 1903,

---

[20] Francis Wilson, 'Unresolved Issues in the South African Economy: Labour', *South African Journal of Economics* 43, no. 4 (1975): 311–27; Dan O'Meara, *Forty Lost Years: The Apartheid State and the Politics of the National Party, 1948–1994* (Johannesburg: Ravan Press, 1996), 173–7, on Anglo-American's sale of General Mining to Sanlam see 120–4; Hermann Giliomee, *The Afrikaners: Biography of a People* (Cape Town: Tafelberg, 2003), 534–47.

[21] Paul N. Edwards and Gabrielle Hecht, 'History and the Technopolitics of Identity: The Case of Apartheid South Africa', *Journal of Southern African Studies* 36, no. 3 (September 2010): 619–39.

[22] Sasha Polakow-Suransky, *The Unspoken Alliance: Israel's Secret Relationship with Apartheid South Africa* (New York: Pantheon, 2010), 73.

but they were only finally built in the early 1970s.[23] And, critically for this story, it was also in 1970 that the Department of the Interior relaunched the population register, powered this time by one of the new IBM computers, a determination to capture all forms of civil registration in a single purpose-built high-rise in Pretoria, and an ominous name.

The Book of Life project was funded on an entirely different scale to the modestly resourced effort that was first run by the Census Bureau. A thirty-story black monolith, called Civitas, was built in Pretoria at the extravagant cost at the time of R10 million, with a special IBM computer room in the basement.[24] And it was staffed lavishly: over 500 clerks and data processors worked in the central office, with one hundred each in the Johannesburg and Cape Town offices, and fifty in Durban. These officials worked exclusively on the population register, with much smaller establishments devoted to the work of issuing passports and registering voters.[25]

The South African IBM project was modelled on the Swedish population register, which from 1947 had used an identity number, a personal paper file and a raft of benefits to support very fine-grained, and largely self-reporting, surveillance of the reproductive, tax and employment histories of every resident.[26] But in one key respect the South African register was different: the architecture of the Swedish system was decentralised, using computer registers after 1967 based, separately, in each of the 300 municipalities. Two historical imperatives drove this decentralisation. The first was the enduring tradition of church-based civil registration in Europe, which made parishes and then municipalities the site of registration, welfare and taxation;[27] and the second was the

[23] W. Willcocks, *Mr Willcocks' Report on Irrigation in South Africa*, BPP Cd 1163 Further Correspondence Relating to Affairs in South Africa (In Continuation of [Cd 903] January 1902) (Daira Sania Co, Egypt, 1901); Tony Emmett and Gerard Hagg, 'Politics of Water Management: The Case of the Orange River Development Project', in *Empowerment through Economic Transformation*, ed. Meshack Khosa (Pretoria: HSRC Press, 2001).

[24] Sue Leeman, 'Sad Saga of the Book of Life', *Pretoria News*, 14 July 1983, sec. 2; Department of Interior, South Africa, *Annual Report for the Calendar Year 1973* (Pretoria, 31 December 1973).

[25] Department of Interior, South Africa, *Annual Report for the Calendar Year 1972* (Pretoria, 1972).

[26] Jack Clarke, interview by Keith Breckenridge and Paul Edwards, 20 April 2004.

[27] Simon Szreter, 'Registration of Identities in Early Modern English Parishes and Amongst the English Overseas', in *Registration and Recognition: Documenting the Person in World History*, ed. Keith Breckenridge and Simon Szreter, Proceedings of the British Academy 182 (Oxford University Press, 2012), 67–92; Paul-André Rosental, 'Civil Status and Identification in Nineteenth-Century France: A Matter of State Control?', in *Registration and Recognition: Documenting the Person in World History*, ed. Keith Breckenridge and Simon Szreter, Proceedings of the British Academy 182 (Oxford University Press, 2012), 137–65.

compelling interest in preventing a repetition of the data-driven genocide of Nazi Germany.[28] Over generations the Swedish state specifically
opted not to allow the development of a single, centralised database of
civil identities.

Centralisation was the object of the South African project.[29] In addition to birth, voting and death certificates that had long been issued
and held in Pretoria, the personal file for each individual in the population register at Civitas contained marriage certificates and gun and
driver's licences that had previously been retained in local magistrates'
and municipal offices. An overweening and naive surveillance goal was
at the heart of the new system. Like the Swedish system, and the Nazi
*Volkskartei*,[30] the population register was supposed to be a continuously
updated register of domicile for all whites, Coloureds and Indians. This
unrealisable surveillance ambition, and the proliferation of linked registration functions, was the Book of Life's undoing. Over time the Book of
Life project would transform into its opposite, becoming a radically simplified biometric register of identification only. That is the plan that has
survived into this century, spreading to Mexico, Brazil, India and many
countries on the African continent.

The project was plagued by unexpected and intractable problems from
the outset. To begin with the construction of the intimidating new sky-
scraper was delayed. The new IBM computers were moved into Civitas
in May 1971, most of the staff took possession of their offices a year later,
but work continued on the building well into 1973. When the department began accepting applications for the new Books of Life in February
1972, they were quickly overwhelmed. Within weeks the department
restricted applications to voters in specific districts and to sixteen-year-
olds who had no other forms of ID. After the excitement of the first year
the project never recovered its momentum; fully six years in, nearly half
of the target population of 7.5 million people had still not received their
new books. As late as 1983, over 100,000 registered (Coloured, Indian
and white) voters had not yet applied for Books of Life.

The new scheme also uncovered (and exacerbated) the regulatory
problems around controlling driving in South Africa. The original plan
was to strip local authorities of the ability to issue licences, primarily for

---

[28] Fred H. Cate, *Privacy in the Information Age* (Washington, DC: Brookings Institution
Press, 1997), 44; IIVRS, *The Impact of Computerization on Population Registration in
Sweden*, Technical Papers (Bethesda: International Institute for Vital Registration and
Statistics, December 1996).
[29] 'Clearing up the Book of Life Confusion', *Rand Daily Mail*, 14 December 1977.
[30] Gotz Aly and Karl Heinz Roth, *The Nazi Census: Identification and Control in the Third
Reich* (Philadelphia: Temple University Press, 2000), 45–55.

the surveillance benefits that would accrue to the central government in Pretoria. But the results were catastrophic. Year after year during the 1970s extensions were required for the compulsory replacement of local licences with the Book of Life.[31] Belatedly the state realised that all black drivers would also have to reapply for new licences.[32] And, in a pattern that, ever since, has threatened the regulatory fabric of daily life in South Africa, the slips of paper that were stuck in to the Book of Life were banally easy to forge, opening the flood-gates to illegal drivers.[33]

The Book of Life also changed the architecture of registration for the much larger population of black people who had fallen under the Bewysburo. Paired with the new centralised population register, the Bantu Homelands Citizenship Act of 1970 set up the process of stripping all black South Africans of their citizenship, allocating them to one of the emerging Bantustans.[34] After the 1976 Soweto Uprising, the state began to use the rhetoric of independence to shift governmental responsibility on to these bantustans. Late in 1977, after representations from the Chief Ministers of the three pseudo-states of Lebowa, Ciskei and Bophutatswana, Vorster announced that all Africans would receive new identification documents 'similar to the Whites' Book of Life' and that entry to and travel in the white territories would be regulated by 'work permits'.[35] At the same time the Department of the Interior began to flesh out a special Kafkaesque nightmare of the bantustans, setting up separate population registers in the three that had been pushed into formal independence,[36] and then building a handful of immigration posts for the 'control of aliens' at arbitrary points on the thousands of kilometres of border linking the bantustans to the old white provinces.[37]

This effort to delegate the problems of consent and government to the bantustans also wrought an important shift in the character of the

---

[31] '3 Million Must Still Get Books of Life', *Pretoria News*, 18 April 1978.

[32] 'Licences Warning to Blacks', *Eastern Province Herald*, 25 January 1978.

[33] 'Rybewyse Is Die Boek Se Swak', *Beeld*, 4 July 1978.

[34] Republic of South Africa, *Bantu Homelands Citizenship Act*, 1970; Liezl Gevers, 'Black Sash and Red Tape: The Plight of the African Aged in KwaZulu and Natal, 1979–1990' (Honours in History, University of KwaZulu-Natal, 2012), 41.

[35] 'Reisdokument Gaan Pasboek Vervang', *Die Volksblad*, 23 July 1977; Jaap Theron, 'Influx Change Expected Soon', *The Citizen*, 3 November 1977; Editor, 'Bewysboekstelsel', *Die Transvaler*, 4 November 1977.

[36] *Transkei. Binnelandse Sake. Bevolkingsregistrasie*, vol. 12/194, BAO, 1968; *Kommissaris-Generaal: Xhosa Volkseenheid: Staatsdepartemente: Republiek Van Suid-Afrika. Justisie Landdroskantore Byhou Van Bevolkingsregister*, vol. 28, KGT, 00000000; *Venda. Bevolkingsregistrasie Sensus En Statistiek*, vol. 11/76, BAO, 1979; *Ciskei. Bevolkingsregistrasie.*, vol. 11/98, BAO, 1981.

[37] Department of Interior, South Africa, *Annual Report for the Calendar Year 1977* (Pretoria, 1977).

overall South African political economy. The budgets of both the fledgling bantustans and those that adopted full independence were increased
dramatically in the last years of the 1970s, with the bulk of the funds
from central government being allocated to health and welfare, especially to the payment of old-age pensions. Mangosutho Buthelezi, the
Chief Minister of KwaZulu, was a particularly skilful manipulator of
the central government's efforts to buy the consent of the people in the
bantustans, accepting funds and adopting bureaucratic resources that
strengthened his authority and withdrawing from those that damaged
it. While he refused to countenance formal independence, by the end of
1978 the government in KwaZulu had issued nearly 1.5 million certificates of citizenship, securing the very large amounts of old-age pension
funding that went with them.[38] It was this shift in the political arrangements of welfare – linking the viability of the bantustans with the distribution of pensions – that set up the movement towards universal biometric
grants a decade later. But, in the meantime, the imperatives of counter-
insurgency prompted another parallel change in the design of the central
South African population register.

## White sacrifice

It was the decision by P.W. Botha's government, in January 1981, to
require fingerprint authentication from all South Africans, white and
black, that created the world's first universal, biometric population
register. Following from that decision, all South Africans were issued
with a common identity document, stripped of many of the surveillance
functions that had originally been included by the document's designers (although the last three digits of each individual's identity number
continued, briefly, to do the critical work of race classification). Three
years later, after flirting unsuccessfully with the idea of decentralising the
population register on the Swedish model, the state began to combine
two massive registers into a single monolith of identity. The first body
of records was taken from the fine-grained biographies of the Book of
Life project in Civitas, and the second from the brutal and inaccurate
fingerprint records that the Bewysburo had stored in an enormous bru-
talist tomb called the New Co-operation Building built in the late 1970s.
The chaotic results of this merged database – with millions of duplicate,
incomplete and inaccurate entries for the black people captured by the
Bewysburo – were exacerbated by the fact that an unknown number of

---

[38] Gevers, 'Black Sash and Red Tape', 41.

the fictional citizens of the Transkei, Bophutatswana, Venda and Ciskei were still excluded.[39]

When the *New York Times* observed in 1981 that the fledgling industry of computerised biometric access control technology was 'a solution to a problem that no one's aware of yet' it was understating the pressure building inside the FBI for an automated solution to the stream of fingerprints.[40] But in South Africa, with a single enormous and unreliable population register as the only recognised source of identification, the urgency for an automated system of biometric identification was already very clear.[41] With the decision to set up universal fingerprint identification as an anti-terrorist strategy the South African government was giving concrete form to the global problem that the newly christened biometrics industry was searching for.

There is a fairly clear correlation between the societies that have moved quickly towards centralised biometric identity registration – Argentina, Chile, Malaysia, Philippines and, more recently, India – and the model of bureaucracy that Stubbs and Rich called the counter-insurgent state.[42] Like these later adoptions of states around the world, the momentum for centralised biometrics in South Africa was also propelled by deeply influential strategies of counter-insurgency. In the weeks preceding the decision to extend fingerprinting to everyone, the African National Congress's special operations unit (under Joe Slovo's command) had staged a series of conspicuous bombings of fuel refineries, banks and railway stations in cities across the country. These attacks coincided with, and reinforced, the frenzied militarisation of the South African state under the Botha government. Key members of the government, like Chris Heunis, the new minister of the Department of the Interior, had begun to set in place a constitutional order called the National Security Management System that redrew the lines of political authority around a set of regional committees dominated by the South African military.[43] The military and its

[39] Valerie Boje, 'How Millions of Files Reflect SA's Reforms', *Pretoria News*, 27 November 1990; 'Zac De Beer: Dr Zackyll and Mr (Take You For A) Ride?', *Noseweek*, July 1993; Tony Stirling, 'Black Sash Claims on Black IDs Repudiated', *The Citizen*, 8 January 1988.

[40] Andrew Pollack, 'Recognizing the Real You', *New York Times*, 24 September 1981; *The Times* was only partly correct, see Simon Cole, 'Digits – Automated Fingerprinting and New Biometric Technologies' (Unpublished paper, n.d.).

[41] Political Staff, 'Why Black Finger Prints?', *Cape Times*, 11 March 1980; John Murray, 'Heunis Keeps a Finger on the Workforce', *The Star*, 5 August 1980.

[42] R. Stubbs, 'The Malayan Emergency and the Development of the Malaysian State', *The Counter-Insurgent State: Guerrilla Warfare and State Building in the Twentieth Century* (1997): 50–71; Paul B. Rich and Richard Stubbs, *The Counter-Insurgent State: Guerrilla Warfare and State Building in the Twentieth Century* (London: Macmillan, 1997).

[43] O'Meara, *Forty Lost Years*, 255–69.

Cabinet supporters – carefully mimicking the arguments of the leading NATO strategist on counter-insurgency – argued that South Africa faced a Soviet-sponsored 'Total Onslaught'.[44] Establishing a national finger-print register of all white, Indian and Coloured South Africans was one of the key elements of the 'Total Strategy' they had designed to pre-serve white power. On the day of the minister's announcement a spokes-man for the South African Defence Force explained that the military believed that it was necessary to 'upgrade the sophistication of identity documents'.[45]

Minister Heunis' announcement was met with a chorus of outrage from legal academics and newspapers, with bitter sarcasm from the far right paper, *Die Afrikaner*, at least one paper pointed to the sad story of the earlier identification campaigns, advising readers to 'stand by for the fiasco'.[46] The minister tried to calm this storm by pointing out that Americans were required to submit to fingerprinting in order to secure their driver's licences. But it was the anti-terrorist argument that ulti-mately prevailed with the NP's supporters. 'There is an onslaught against the Republic', the Cape paper *Die Burgher* explained, 'and it is neces-sary to prevent terrorists and other enemies of the country from using false identity documents, drivers' licenses and other documents in order to conduct espionage and sabotage against the country'.[47] In the years immediately following the proposal of the universal fingerprint ID the state was preoccupied with the bizarre constitutional re-engineering of Heunis' Tricameral Parliament (which established a carefully gerry-mandered separate parliamentary chamber and cabinet for each of the white, coloured and Indian racial electorates), and the introduction of the fingerprinted ID book only came into effect five years later. In an interesting example of the ways in which information infrastructures can become invisible, most South Africans know this ID Book, not as

[44] Philip H. Frankel, *Pretoria's Praetorians: Civil-Military Relations In South Africa* (Cambridge University Press, 1984), 46–70; Mark Swilling and Mark Phillips, 'State Power in the 1980s: From "Total Strategy" to "Counter-Revolutionary Warfare"', in *War and Society*, ed. Jacklyn Cock and Laurie Nathan (Cape Town: David Philip, 1989).

[45] Ameen Akhalwaya, 'Storm over Govt's "Pass" Plan for All', *Rand Daily Mail*, 15 January 1981.

[46] Editorial, 'Do We Need It?', *The Natal Mercury*, 16 January 1981; Editorial, 'It's a Blot', *The Citizen*, 17 January 1981; 'Vingerafdrukke', *Die Afrikaner*, 23 January 1981; 'Now Stand by for South Africa's Great Fingerprint Fiasco', *Sunday Express*, 18 January 1981.

[47] 'Daar is 'n aanslag teen die Republiek en dit is noodsaaklik dat terroriste en ander vyande van die land verhinder word om valse identiteitsdokumente, rybewyse en ander dokumente te gebruik om op die land te spioeneer of sabotasie te pleeg.' 'Vingerafdrukke Gevra', *Die Burger*, 15 January 1981.

a fingerprinted document, but as the 'barcode ID' because it presents a machine-readable strip on its opening page.[48]

The first draft of the new Identification Bill was presented to the white House of Delegates just days before the second, and indefinite, nation-wide State of Emergency was announced by P.W. Botha.[49] At a time when early morning Security Police raids were becoming common, criticism of the bill was much more measured than the responses to Heunis' original proposal. In the place of the dismissive and sarcastic assessments of the earlier debate, the liberal papers were left asking the same plaintive question they addressed to everything else that Botha did: is this really reform?[50] With the argument about national security completely persuasive, English and Afrikaans speakers were reluctantly prepared to pay the price for a properly working fingerprint surveillance system for Africans.[51] In the second reading of the bill the minister agreed to change the (impossible) requirement that all South Africans be fingerprinted immediately to a stipulation that the registrations be complete within five years. The combination of this false concession and the removal of racial identification from the ID book left the liberal press searching for a political silver lining on the dark cloud of universal fingerprint registration; 'it at least makes for a kind of justice', the editor of the *Daily News* observed, 'that all South Africans of all races will carry the same kind of ID books'.[52] For black South Africans, the debate amongst mostly English-speaking whites about the loss of privacy and dignity was insulting. 'Whites have just realised that fingerprinting civilians is humiliating and associated with criminality', the editor of the *Sowetan* commented, 'while generations of blacks have gone through the experience almost routinely'.[53]

It is interesting that the Afrikaans press, like Gandhi almost a century earlier, presented the forfeiting of white personal liberties as a noble sacrifice in the effort to secure the workings of the national fingerprint register.[54] There was, officials argued, something righteous in white, Coloured and Indian people's adoption of the same regime of identification that had been directed at Africans. 'The black people in South Africa have

---

[48] Bowker and Starr, *Sorting Things Out*.
[49] O'Meara, *Forty Lost Years*, 343.
[50] 'Fingerprinting SA', *The Cape Times*, 8 April 1986.
[51] E. Brits, 'Kwessie van Ras en Vingerafdrukke Pla Nog', *Die Burger*, 15 June 1990; 'Almal Gelyk Met Vingerafdrukke', *Die Vaderland*, 2 July 1986; 'Why Fuss about Fingerprinting?', *The Natal Mercury*, 26 April 1986.
[52] 'Good Move If', *Daily News*, 9 November 1984.
[53] 'Comment', *Sowetan*, 4 April 1986.
[54] Vaderlandspan, 'Bewysboek Verwelkom', *Die Vaderland*, 15 January 1981.

always been subject to fingerprinting', the Director-General of Home
Affairs, Gerrie van Zyl, explained. 'This is now a case of the whites,
Asians and Coloureds being placed on the same standing with the blacks
in order to build up a national fingerprint register for the whole country.'[55] Under P.W. Botha's Presidential Council government, the finer
moral points of the parliamentary debate were, in any case, irrelevant. As
if to demonstrate that the decision to fingerprint everyone now lay in the
hands of the officials South Africans called the securocrats, Home Affairs
began collecting fingerprints before the bill had even been approved.[56]

The most important effects of the decision to introduce the single ID
book were felt in the size and organisation of the fingerprint authenticated population register owned by Home Affairs. The issuing of
the new ID coincided with the final dismantling of the Native Affairs
Department (now operating under the Orwellian title of the Department
of Cooperation and Development), and Home Affairs inherited the key
registration functions of Hendrik Verwoerd's old department.[57] A collection of sixteen million sets of fingerprints, built up from the issuing of the
Dompas by the old Central Reference Bureau, formed the foundation of
the new population register. Home Affairs took over the fingerprint collections housed in the massive Cooperation Building, built for Heunis'
department in 1981.[58] Faced with the management of this huge collection of fingerprints, and with an almost infinite requirement for skilled
fingerprint readers, Home Affairs searched for an automated remedy,
without success.[59]

The new identity books, and the abolition of influx control that coincided with them, dramatically changed the ordeal of identification and
movement for Africans who could not show permanent urban residence
in the cities.[60] They were not easily available to people who had Reference
Books or birth certificates from one of the four nominally independent
homelands but, in the two years after 1986, five million Africans swopped
their Reference Books for the new IDs, an average of some 20,000 applications per day. By the end of the decade Home Affairs' Fingerprint
Section was operating at a frenzied pace: in a single year the division
classified 1.5 million new sets of prints, identified 160,000 people for the

---

[55] 'Almal Gelyk Met Vingerafdrukke'.
[56] 'Let the House Ponder; Fingerprints Are Here', *Weekly Mail*, 22 May 1986.
[57] Brian Pottinger, 'The Fall of an Empire That Ruled Black Lives', *Sunday Times*, 10
November 1985.
[58] 'Zac De Beer: Dr Zackyll and Mr (Take You For A) Ride?'.
[59] Jonathan Crush, 'Power and Surveillance on the South African Gold Mines', *Journal of
Southern African Studies* 18, no. 4 (1992); 'HANIS Basic System Commissioning', 18
February 2002, www.info.gov.za/speeches/2002/0202191046a1001.htm.
[60] Stirling, 'Black Sash Claims on Black IDs Repudiated'.

Justice System, and conducted over two million confirmation of identity searches. By 1991, Home Affairs had captured the fingerprints of another nine million people, leaving them without 'funds, manpower or accommodation' to carry on the task.[61]

Many white people, especially those who had the original Book of Life issued in the 1970s, did not apply for the new, fingerprinted, identity document until the deadline approached in 1990, the same year that the African National Congress was unbanned. There is, I think, some irony in the fact that thousands of white South Africans queued to be fingerprinted for ID books designed as an anti-terrorist measure, and instrument of white racial supremacy, in the year after Nelson Mandela was released from prison. By this time officials had begun to stress different reasons for holding the books, like the usefulness of fingerprints for 'orderly public administration', for business or for identifying dead bodies, but the desire for closure, for completing the national register, had now become both the end and the means of the state's policy.[62]

After 1986 the completion of the centralised biometric population register became an always receding goal, but South Africans have remained doggedly on the road to it. Even as the last of the original, racially-defined, Books of Life were being replaced, some South African businesses began to lobby for the replacement of the new ID book, with its separate pages for voting records, gun and vehicle licences with a more convenient, single-purpose, identity card – the format that had originally been issued to whites during the 1950s, but, this time, with biometrics as the single leash of authentication. When *The Citizen* suggested that one of the reasons for a new card was that the old book was too easily 'subject to counterfeiting', the outraged Acting Director-General of Home Affairs responded that 'the document at present in use, with laminated material over both the personal and driver's particulars, is for all practical purposes impossible to counterfeit'.[63] A decade later the department would eat those words.

The struggle to complete the central biometric register – begun in the 1950s as an alternative to literate civil registration for Africans in the countryside and extended to all by the national security ambitions of the 1980s – is still under way today. In recent years in the face of a

---

[61] Staff reporter, 'Rush for New Non-Racial ID Documents', *The Star*, 7 January 1987; quote from Staff reporter, '8-M Blacks Get Identity Docs', *The Citizen*, 18 February 1991.

[62] Brits, 'Kwessie van Ras en Vingerafdrukke Pla Nog'; Stirling, 'Black Sash Claims on Black IDs Repudiated'.

[63] Keith Abendroth, 'Non-Racial ID Document Is Considered', *The Citizen*, 15 October 1990; Act Director-General, Department of Home Affairs, 'Not Racial ID', 1 November 1990.

proliferation of biometric demands and functions – in policing, immigration, banking and welfare – the democratic state has worked assiduously to reinvent the central population register and issue a new biometric identity card.[64] But the struggle to make the central register work remains only half of the story of the South African biometric state. For the failure of the central government's national security project has been matched by a proliferation of privatised biometric systems. And the most important of these – which has served as both the model and vehicle for the adoption of an infrastructure of biometric citizenship derived from South Africa around the former colonial world – was the development of the system of biometric social grants. That story has its roots in Zululand in the middle of the 1980s.

## Biometric banking in the Bantustans

The story of the development of biometric social grants in South Africa is an intriguing combination of feminist politics, indirect rule and banking. Money for the grants stemmed from the indispensability of pension payments to the Apartheid state's effort to delegate control over the rural poor to the Bantustans. After 1976 the central government began to increase the funds for black pensioners. But the organisational pressure to find a practical means of delivering the pensions, in the absence of even the most basic administrative and transport infrastructure in the countryside, emerged from the Black Sash, a movement of liberal, middle-class feminists skilled in the use of the law against the Apartheid state. The themes and patterns of this movement – focusing on the effects of bureaucratic incompetence and corruption on the lives of deserving poor women – bear a striking similarity to the ill-fated history of what Skocpol called the maternalist welfare state in the United States.[65] A decade later, it was this same social movement of women, and a loose alliance of feminist experts within it, that pushed through the development of the universal Child Support Grant in South Africa.[66] The passing of the Welfare Laws Amendment Act in 1997, and the universal biometric grants it allowed, can be said to mark the completion of the South African project of biometric citizenship.

[64] For this story, see Keith Breckenridge, 'The Elusive Panopticon: The HANIS Project and the Politics of Standards in South Africa', in *Playing the ID Card: Surveillance, Security and Identity in Global Perspective*, ed. Colin Bennett and David Lyon (London: Routledge, 2008), 39–56.

[65] Theda Skocpol, *Protecting Soldiers and Mothers: The Political Origins of Social Policy in the United States* (Cambridge, MA: Harvard University Press, 1992), 2.

[66] For a narrative account of the passing of the Act, see Francie Lund, *Changing Social Policy: The Child Support Grant in South Africa* (Pretoria: HSRC Press, 2007).

If the legitimacy problems of the Apartheid state and feminist welfare reform presented two necessary causes of the final episode of the
South African biometric state, a third enabling agent of the project was
First National Bank (FNB), the disinvested local remnant of the British
Barclays Bank and one of the four very large South African finance
houses. FNB, under conspicuously liberal management, had its own
interest in developing information technologies that would equip it to
capture new customers from the millions of the unbanked poor. The
relationships between each of these agents were often difficult, with each
often in fierce public conflict with the other. Yet, ironically, it was this
dispersal of interests and objectives – from the early conflict between
the Black Sash and the KwaZulu bantustan government – that drove the
momentum of biometric citizenship.

Over the course of the 1970s the South African government carved up
the territory of the eastern half of the country in the effort to create ten
nominally independent countries. The original idea behind the strange
patchwork of territories (as I explained in the previous chapter) was to
mimic the process of decolonisation that was signalled by Macmillan's
'Winds of Change' speech in 1960; Prime Minister Verwoerd and some
of his key advisors were firmly committed to the idea that each Bantustan
was a reflection of the politics of ethnic self-determination, each accommodating a separate ethno-linguistic tribe.[67] In practice the demography
and territory of the Bantustans were much more pragmatically part of
the effort to sustain white supremacy. Borders were drawn to mop up
significant numbers of black people and (most of the time) to exclude
geographical or economic resources. Leaders of these ridiculous homeland governments were almost all selected for their compliance with the
demands of the central government (even where, as was often the case,
they were opposed by conservative and patriarchal local leaders).[68]

Mangosutho Buthelezi, chief minister in KwaZulu, was an exception to the general rule of craven bantustan leadership. The brutality of
the conflict between Buthelezi's Inkatha Freedom Party and the UDF/
ANC in Natal in the 1980s and 1990s has mostly obscured the fact that
Buthelezi was for many decades an effective opponent of the Apartheid
state. From the middle of the 1970s he fought an intriguing battle for
autonomy in Zululand while insisting that he, and Zulu-speaking people
in general, were full citizens of a unified South Africa. He was a master of institutional passive-aggression, withholding his participation from

---

[67] Giliomee, *The Afrikaners*.
[68] Peter Delius, *A Lion Amongst the Cattle: Reconstruction and Resistance in the Northern Transvaal* (Oxford: James Currey, 1996).

the most daft and contaminating constitutional arrangements of the Botha state, whilst carefully assembling the administrative resources of provincialism.[69]

For the central state, welfare spending during the 1970s worked as one of the very few effective instruments for bolstering the absurd sovereignty of the Bantustans. After accepting self-governing status in 1977, KwaZulu was rewarded with control over the distribution of pensions and a significantly increased budget in 1979 of over R204 million, about a third of which was allocated to the combined health and welfare department.[70] All women over the age of sixty, and all men over sixty-five, were entitled to a payment of R66 every two months. But with the offices of the Department of Health and Welfare based in the homeland capital of Ulundi – over three hours' drive on difficult roads from the largest centres of population around the cities of Durban and Pietermaritzburg, and with no meaningful administrative infrastructure at all, the entitlements of the pensions system soon became a source of bitter discontent.

The Durban offices of the Black Sash quickly became the focal point for protests about the failings of the pension system in KwaZulu. The themes of administrative incompetence, bureaucratic arrogance and indifference to the suffering of old women structured the Black Sash campaign from the outset. As early as March 1980 Jill Nicholson, a Durban official, reported to the organisation's annual conference that pensioners faced a delay of more than a year after registering for their pensions, and she described – in terms that made for excellent newspaper copy – a visit to a pay point where pensioners who had been waiting for three days confronted a payout officer who spent only three hours at his post and 'constantly shouted at people and berated them'.[71]

Numerically the Black Sash was a tiny political organisation. For its influence it had to rely, in part, on the press eager for stories about corruption and administrative brutality. But it also made clever use of the law, working closely with the small but effective fraternity of labour lawyers sustained by the growing union movement.[72] It was this cunning use of lawsuits against the KwaZulu government that converted the pension

---

[69] Theron, 'Influx Change Expected Soon'; Editor, 'Bewysboekstelsel'; D. Glaser, 'Behind the Indaba: The Making of the KwaNatal Option', *Transformation* no. 2 (1986), http://transformation.ukzn.ac.za/index.php/transformation/article/viewFile/394/213.

[70] African Affairs Correspondent, 'Health and Welfare Get Priority in KwaZulu', *The Natal Mercury*, 29 May 1979; Gevers, 'Black Sash and Red Tape', 1.

[71] Labour Correspondent, 'Survival Is Incidental to These "Shuffling Bureaucrats"', *Rand Daily Mail*, 14 March 1980; Editor, 'Black Pensions', *The Daily News*, 4 September 1982.

[72] Gevers, 'Black Sash and Red Tape', 47–9.

funds distributed from the centre from a slush-fund for sustaining an illegitimate bureaucracy into a meaningful entitlement. In a well-known speech in 1983, the KwaZulu Auditor-General admitted that the fiscal burden of pensions had been transformed by the legal battles of the Black Sash. No longer could the arbitrary withholding of payment to individuals serve as a 'quick way of saving' for the bantustan fiscus. 'Unfortunately the pensioners are now aware of their rights', he explained to the legislature, 'thanks to Legal Aid and the Black Sash.'[73]

In fact, of course, the KwaZulu government's attitude to welfare funding, and to the Black Sash's campaign to secure the rights of individual pensioners, was more ambivalent. Buthelezi was furious about the litany of accusations of corruption and abuse that provided the energy and outrage of the Black Sash campaign. But he also, early on, recognised that pensions were potentially invaluable tools for injecting cash into the desolate economies of rural Zululand. He duly endorsed the Black Sash argument that the central government was 'legally bound to pay pensions to every South African' and that the funding going to KwaZulu to meet these demands needed to be increased, exponentially, to match the rapidly growing population.[74] As always he looked for ways to bolster his own provincial power. Early in 1984, as powerful white business interests in Natal and even key figures in the National Party began to take seriously the possibility of secession by a combined KwaZulu-Natal within a federalist South Africa, Buthelezi appointed a commission of enquiry, chaired by an eminent economist at the University of Natal with strong links to the Black Sash, into the standardisation and regularisation of the payment of pensions in KwaZulu.[75]

It was the report of the Nattrass Commission – immediately, and very publicly, endorsed by the KwaZulu government[76] – which drew the blueprint for the new system of pension payments that was developing in the late 1980s. Nattrass emphasised three basic points about pensions that equipped the bantustans with a powerful lever over the central government fiscus. Pensions were, in the first instance, an inalienable right of all South Africans, regardless of whether they were residents in KwaZulu (or any other bantustan). Second, the differentiation of pension levels for

[73] Quoted in Francie Lund, 'Children, Citizenship and Child Support: The Child Support Grant in Post-Apartheid South Africa', in *Registration and Recognition: Documenting the Person in World History*, ed. Keith Breckenridge and Simon Szreter, Proceedings of the British Academy 182 (Oxford University Press, 2012), 480.

[74] Sowetan Correspondent, 'Pensions Backlog: KwaZulu's Old Age Budget Has Dried Up', *The Sowetan*, 23 January 1984.

[75] Glaser, 'Behind the Indaba'; 'Zulus Appoint Probe into Pension Schemes', *The Daily News*, 11 April 1984.

[76] Daily News Reporter, 'KwaZulu Accepts Pension Proposals', *Daily News*, 7 May 1986.

whites, Coloureds and Africans – which had been falling steadily since the early 1970s – should be abandoned. And then, third, the central government was responsible for financing the payments. Nattrass also insisted on two practical conditions for the delivery of payments: the creation of a network of convenient payment points to replace the chaotic and unreliable system of temporary payment offices and, importantly, the development of an accurate system of records.

A year later the KwaZulu government appointed First National Bank (hereafter FNB) to do the difficult administrative work of registering, accounting and delivering pensions in the rural backwoods. FNB formed a company called Cash Paymaster Services (CPS), which ever since has dominated the field of privatised social grant payments. The CPS contract broke down to three distinct tasks. The first was nominating unique recipients on the basis of data that was received from the KwaZulu Department of Welfare and Pensions, the second was building a record system to track payments and the third was finding a means of delivering cash to the 300,000 people who lived in the inaccessible valleys of the KwaZulu hinterland.

The company that took on the technical work for CPS was called Datakor. The firm existed for less than a decade, the local rump of the multinational Unisys, which disinvested from its South African operations in 1988 and then bought them back in 1995.[77] Unisys was itself the result of the merger, in 1986, of the venerable military contractor Sperry Rand and Burroughs, a manufacturer of office and banking equipment. By the end of the 1980s Unisys had begun to specialise in the development of computer systems for the needs of local government in the United States, using standard IBM-compatible hardware and Microsoft operating systems, especially in the management of welfare payments and health care.[78] And it was Unisys, after 1993, that carried the biometric identification and cash transfer technologies away from South Africa. In KwaZulu, Datakor's work on banking systems for delivering cash to ATMs, and keeping electronic records of payments, provided half of the solution that Buthelezi's government was looking for. But the problem of registering and identifying the recipients of pensions in the countryside remained unsolved for several years.

By the end of 1988 the goodwill that had followed the publication of the Nattrass report had dissipated. The Black Sash began to complain,

---

[77] Simon Cashmore, 'Datakor: Digesting Unisys', *Financial Mail*, 17 August 1990.
[78] G.T. Gray and R.Q. Smith, 'Against the Current: The Sperry-Burroughs Merger and the Unisys Struggle to Survive 1980–2001', *Annals of the History of Computing, IEEE* 29, no. 2 (June 2007): 12–13.

once again, of brutally long queues, of old people having to sleep at pay points to have any chance of getting their pensions. And they lobbied the KwaZulu government to introduce regular monthly payments to replace the erratic bi-monthly cycle.[79] In its annual report the organisation reminded the KwaZulu government that 'five years have passed' since the Nattrass Commission, but that very few of the recommendations had actually been implemented.[80]

In the background Datakor (with FNB's private customers in mind) was experimenting frantically with an expensive PC-based voice-recognition system hooked up to an ATM on the back of a four-wheel drive pick-up truck. The initial reports on these voice-recognition tests were very optimistic. Voice recognition was 96 per cent accurate, and it seemed only to fail when the subjects were intoxicated or under acute stress. It was, also, much faster than automated fingerprinting.[81]

But as the tests grew in scale the reliability of voice recognition collapsed, and Datakor was forced to fall back on the use of fingerprints.[82] As was so often the case in the long history of fingerprinting in South Africa, it was the registration requirements of the mining industry which again underpinned the adoption of fingerprints. From the early 1980s, the mines had been investing heavily in the development of automated systems for their enormous collections of fingerprints.[83] At almost exactly the same time as FNB was awarded the KwaZulu pensions system, Datakor had won a tender for converting the Chamber of Mines' manual photographic and fingerprinting systems for the registration of contracts to a computerised, magnetic card ID. The new system was designed to link the recruiting headquarters on the south side of Johannesburg to over one hundred recruiting offices scattered through the most dishevelled towns and villages in the country, and to the dozens of mines along the Witwatersrand. This project meant that, at the same time as the KwaZulu pension system was being developed, Datakor was busy with the registration of hundreds of thousands of mine workers on the different mines using mobile digital fingerprinting units.[84]

And the fingerprint biometrics worked. As FNB explained in its successful application for the 1995 Smithsonian Computerworld Award for innovation in IT, the technology they were using was 'not at the leading

[79] 'Bid to Pay Pensions Monthly', *The Natal Mercury*, 29 March 1989.
[80] Daily News Reporter, 'Black Sash Slams KwaZulu Government's Pension Scheme', *Daily News*, 30 September 1989.
[81] Finance Editor, 'KwaZulu System Sign of the Future', *Natal Mercury*, 18 July 1990.
[82] Editorial, 'Pensions: Biometric Answers', *Financial Mail*, 5 June 1992.
[83] Crush, 'Power and Surveillance on the South African Gold Mines', 840.
[84] *Ibid.*, 840–3.

edge'. Biometric scanners were already being used for access control in many companies. And all the components used in the KwaZulu project were conventional off-the-shelf products – commercial biometric scanners made by Identix and Pentium 486 PCs running MS-Dos. The technical description that FNB submitted with its explanation to the Smithsonian admitted that they really had no idea if 'on a worldwide basis ... this approach to delivering "aid" is unique or not'. The only real technological modifications were reducing the matching requirements of a single fingerprint to accommodate real fingers and their prints, adding additional fingerprint templates for each customer's user profile and segmenting the central database to speed up matching.

But the new system surprised its developers. Even the fear of fingerprinting – and its associations with the Dompas – which had prompted FNB to try voice recognition in the first place, dissipated rapidly, overcome by the fact that fingerprint-authenticated pensions were, as the bank put it (unconsciously echoing Agamben), 'the only form of sustainment of life'.[85] By the end of 1992 over 100,000 pensioners were receiving biometric payments from FNB, by 1995 the figure had reached 320,000. And the scheme had begun to spread by imitation when the Cape provincial government adopted the same technology for its own, mainly urban, pension payments system.[86]

From the beginning, the media enthusiasm that was generated by the South African biometric pensions system – unlike the grim welfare arrangements developed by the Clinton administration shortly afterwards in the United States[87] – rose from the promise it offered of delivering financial products to the billions of people locked out of the banking system by the technologies of writing. It was the banking opportunity of the FNB project that was clear from the outset: nearly twenty-five million people 'have no bank account at all', the bank explained, because 'many are functionally illiterate'.[88] Using biometrics to escape the constraints of literacy for identification and transactions in combination with the secure off-line record keeping on smart-card chips opened up the promise of banking and especially credit products to hundreds of

---

[85] Cash Paymaster Services, 'Questionnaire: Computerworld Smithsonian Awards' (Computerworld Smithsonian Awards, 1995), Archives Center at American History, Smithsonian Institution; Giorgio Agamben, *Homo Sacer: Sovereign Power and Bare Life* (Berkeley: University of California Press, 1998).

[86] SAPA, 'CPA Fingerprint Pension Move', *The Citizen*, 16 November 1993.

[87] J.J. Killerlane III, 'Finger Imaging: A 21st Century Solution to Welfare Fraud at Our Fingertips', *Fordham Urban Law Journal* 22 (1994): 1327; Smith, *Welfare Reform and Sexual Regulation*.

[88] Cash Paymaster Services, 'Questionnaire: Computerworld Smithsonian Awards'.

millions of new customers on the continent. And international journalists quickly picked up this theme from the managers at FNB (and other South African banks). The South African pensions scheme held out the prospect of leap-frogging the difficult communications of conventional banking to reach the very poorest around the world.[89] This desire to capture the unbanked remains one of the most powerful drivers behind the proliferation of biometric financial services in the former colonial world.

### The Child Support Grant

The other motivation behind biometric citizenship follows from a global turn, dating from the middle of the 1990s, to the use of direct cash payments to address the subsistence crises faced by poor families. This is an international policy movement, based in middle-income former colonies like Brazil, Indonesia and India.[90] The Mexican system, called *Oportunidades*, imposes harsh conditions on recipients and allows for only three years of support; it was developed with the support of the World Bank, and it is sometimes presented as the fountainhead of what has since become a global movement.[91] By comparison the South African system of grants is much larger, practically unconditional and indefinite in its duration – all of which make it little favoured by the World Bank. Both projects were conceived in 1996, and the Mexican grants began as a pilot in August of 1997. The South African grants were first paid out eight months later. Unlike the Mexican system, however, they were delivered nationally and relied heavily on biometric registration and payment from the beginning. It was only much later, in 2010, that Mexico began to introduce biometrically authenticated electronic payments.[92]

---

[89] Paul Penrose, 'Out of the Backwoods', *The Times*, 10 February 1995; Mark Ashurst, 'First World Smartcards and Third World Pensioners', *Financial Times (London)*, 16 February 1996; Jeffrey Krasner, 'S Africa on Cutting Edge of Bank Tech', *Boston Herald*, 19 August 1995; 'SBSA – Banking the Unbanked', *Computerworld Honors Program*, 1998, www.cwheroes.org/Search/his_4a_detail.asp?id=3453; Mark Ashurst, 'South Africa's Banking Revolution: South Africa's Remote Rural Regions Have Become an Important Testing Ground for New Banking Technology, such as Mobile ATMs and Biometrics', *The Banker*, 1 October 1998.

[90] Hanlon *et al.*, *Just Give Money to the Poor*.

[91] S. Levy, *Progress against Poverty: Sustaining Mexico's Progresa-Oportunidades Program* (Washington, DC: Brookings Institution Press, 2006), 114; Hanlon *et al.*, *Just Give Money to the Poor*, 130–1.

[92] Arteaga, 'Biological and Political Identity'; The Fletcher School, Tufts University and Bankable Frontier Associates, 'Is Grandma Ready for This? Mexico Kills Cash-Based Pensions and Welfare by 2012'.

The combination of direct cash payments for the poor and biometric systems of delivery has quite distinctive roots in South Africa.

Yet it is also true that the policy-makers behind the South Africa Child Support Grant had in mind very few of the moralising objectives that preoccupied their Mexican contemporaries. In place of the statistical obsessions with improvements in school attendance, diet and participation in the formal health-care system, the South Africans were concerned mainly to iron out the strange racist inconsistencies of the grants paid by the Apartheid state and, critically, to avoid the financial implications of extending the generous benefits developed for whites to the entire population. It was that cost-saving imperative – fostered by a parsimonious National Treasury – that accounted for the astonishing speed of the Lund Commission's deliberations and law-making. Another important part of the South African welfare state was – as it was a century earlier in the United States[93] – the flexibility and uncertainty created by the opening of a new political dispensation.[94] On 1 April 1998 – just two years after the appointment of the commission – the grants began to be paid.[95]

The feminist roots of the Child Support Grant were most obviously visible in the selection of beneficiaries: here the committee resolved to direct financial aid to the person primarily responsible for the child, regardless of that person's marital status or their biological relationship with the child. The very heavily gendered character of child-care work in South Africa has meant each year – as the eligibility of children and numbers of recipients have grown – a very significant fiscal transfer to women, young and old. As of 2012, 99 per cent of the seven million Child Support Grant recipients were female, which, when it is combined with the 70 per cent of state pension recipients who are women, represents something like a revolution in the basic structure of the regional economy.[96] What is particularly striking about this feminist

---

[93] Skocpol, *Protecting Soldiers and Mothers*.

[94] Lund, 'Children, Citizenship and Child Support', 491; Shireen Hassim, '"A Conspiracy of Women": The Women's Movement in South Africa's Transition to Democracy', *Social Research* 69, no. 3 (Fall 2002): 693–732.

[95] Lund, *Changing Social Policy*, 65.

[96] Progressus Research Development Consultancy, *The Payment Experience of Social Grant Beneficiaries* (FinMark Trust, April 2012), 131. This shift in the flow of resources in the economy has been intensified by the entry of one million women into the labour market after the removal of influx controls in 1986, a change that is often, and incorrectly, described as a crisis of employment in general. Abhijit Banerjee *et al.*, *Why Has Unemployment Risen in the New South Africa*, Working Paper (National Bureau of Economic Research, June 2007), 17–18, www.nber.org/papers/w13167; H. Bhorat and M. Oosthuizen, 'Employment Shifts and the "Jobless Growth" Debate', in *Human Resources Development Review* (Pretoria: HSRC Press, 2008).

coup is that – while it reflected the determination and capacity of women's organisations within the democratic movement[97] – it was contradicted, largely ineffectually to date, by a broadly conservative view of the patriarchal family within the ANC.[98]

Another paradoxical outcome emerged from the selection of the delivery systems for the Child Support Grant. The Lund Commission, drawing on the history of resisting the Thatcherite reforms in the 1980s, had opposed the idea of using private firms to deliver the grants, worrying, with good grounds, that the provincial governments lacked the ability to regulate the contracts and work of companies like Cash Paymaster Services. But the efficiencies of electronic banking, and of CPS's biometric systems in particular, have proven indispensable to the expansion of the grants, allowing the state to expand the pool of beneficiaries from zero to all children under seven years between 1998 and 2002, and then, to all children under fourteen years, in the years thereafter. Fifteen years after the introduction of the grant, eighteen million people (one-third of the total population) were receiving monthly payments – this massive expansion of privatised welfare capacity came at a time when the state was otherwise struggling to bring about change in the work of the national bureaucracy.[99]

Over the last fifteen years Cash Paymaster Services has stuck doggedly to its knitting, carefully fostering the technological momentum of biometric registration and financial services. Since 1992 the ownership of the company has changed significantly, with FNB selling its holdings to Serge Belamant's Net1 UEPS. Yet the company has remained the primary advocate and beneficiary of the argument that the delivery of grants to poor South Africans requires fingerprint biometrics. Through this period CPS has resisted some of the grand ambitions of biometric state-building. When the Department of Home Affairs looked to build a biometric panopticon – a single smart-card system that would coordinate the population register with banking, health care, vehicle licensing and the justice system – CPS refused to participate, diligently building up its own, private, infrastructure in the countryside for grant payments and credit.[100] After a decade of confusion the Home Affairs project subsided

[97] Hassim, 'A Conspiracy of Women'.
[98] Shireen Hassim, 'Social Justice, Care and Developmental Welfare in South Africa: A Capabilities Perspective', *Social Dynamics* 34, no. 2 (2008): 110; South African Government, 'Green Paper on Families: Promoting Family Life and Strengthening Families in South Africa', 12 August 2012, www.thepresidency.gov.za/pebble. asp?relid=6654, especially 9–10 which calls for a turn to the Latin American focus on families as recipients of grants.
[99] Breckenridge, 'The Biometric State'.
[100] *Ibid.*; Breckenridge, 'The Elusive Panopticon'.

to a chastened single-purpose identity card, and CPS was left in possession of the biometric payment infrastructure (an advantage that it was able to use to eliminate its competitors from the South African banks in a high stakes tender struggle in 2012).[101]

Another intriguing characteristic of the development of the biometric grant system is that its rise coincided with an increasingly strident public discourse about welfare corruption. The early public criticisms from the Black Sash about the inefficiencies and brutality of the KwaZulu pensions system had been generally free of references to fraud, either on the part of the officials in the system or of the beneficiaries. In the Nattrass report, commissioners had worried that irregular payment points encouraged recipients to claim multiple pensions at different sites and they pointed to the likelihood that the unpredictable manual allocation of grants (which frequently resulted in large numbers of people being turned away after being told that 'there is nothing for you') facilitated rent-seeking from officials.[102] But by the early 1990s, as the new computer systems were being implemented, the Black Sash was beginning to protest loudly against the pervasiveness of embezzlement and illegal forms of gatekeeping.[103]

So it was that when the new biometric systems were announced they had adopted pension fraud as their reason for being. Key to these arguments was the claim that the adoption of the expensive systems of biometric payment was covered by the elimination of invisible pilferage of 'millions of rands a year'.[104] This argument – that expensive private biometric systems pay for themselves out of what is commonly called leakage – also now works as a global justification for biometric administration. In South Africa it had some grotesque local features derived from the long history of mimetic registration. The most damaging of the many public protests about fraud in the pension system – which echoed through public debates for years doing untold harm to the national reputation of the state bureaucracy – was the scandal of officials in the former Ciskei using their penises and toes to create inked fingerprints on false

---

[101] Andisiwe Makinana, 'Social Grants: 15.3-Million People to Register Again – News – Mail & Guardian Online', *Mail and Guardian*, 15 February 2012, http://mg.co.za/article/2012-02-15-social-grants-153million-people-to-register-again; Piet Rampedi, 'Battle for R10bn Tender', *The Star*, 29 May 2012; Amanda Visser, '"No Irregularity" in R10bn Cash Paymaster Tender Win', *Business Day Live*, 28 March 2013, www.bdlive.co.za/national/2013/03/28/no-irregularity-in-r10bn-cash-paymaster-tender-win.
[102] KwaZulu, *Report of the Committee of Enquiry into the Payment of Social Pensions in KwaZulu*, 1985, 70, 90.
[103] 'Pensioners Asked to Pay "Special Tax"', *The New Nation*, 9 May 1991.
[104] Editorial, 'Pensions: Biometric Answers'.

pension claims.[105] Widespread official fraud now replaced inefficiencies and accessibility as the primary problems of the old paper-based systems of payment. This argument allowed both the companies and the ministers who appointed them to present the counter-intuitive case that the expensive new systems 'saved millions of rands'.[106]

In practice the actual story of corruption and theft was more arcane. Most noticeable was a dramatic increase in the number of very violent cash-in-transit heists, as gangs armed with assault rifles fell on the little convoys carrying the cash-dispensers and their mobile millions along the dilapidated roads of the bantustans.[107] Over time the state shifted more and more of the responsibility for securing the payment sites to the companies responsible for delivering the grants. But a much more significant shift in the public language of corruption emerged from the decision-making process around the very profitable social grant tenders. In general the firms responsible for pension transfers extract a fee of about 10 per cent of the total social grant budget. This has always meant that the transfer tenders were very profitable businesses, but in recent years, as the population of grant recipients has grown from hundreds of thousands to tens of millions, the tender values have grown enormously: in 2012 the fee income on the single national tender amounted to more than R10 billion, by far the single largest government tender.[108]

From the earliest provincial contracts to the current national scheme involving the registration of nearly twenty million people, the tenders for the provision of biometric social grants have been the focus of intense public controversy. Often unsubstantiated accusations that the private grant providers have bribed key decision-makers in the state lie at the heart of all of these disputes. Some of them, like the prosecution and imprisonment of the National Minister for Social Welfare, Abe Williams, in 1996, have been proven.[109] Often, however, the accusations live on in the netherworld of rumour, with information being provided by

---

[105] Nick Grubb, 'Below-the-Belt Scam Uncovered', *Saturday Star*, 16 November 1996; David Macgregor, 'Penis Print Scam Bust', *Sunday Independent*, 8 June 1997.

[106] Cash Paymaster Services, 'Questionnaire: Computerworld Smithsonian Awards'; see also Sy Goodman, 'Computing in South Africa; an End to Apartness?', *Communications of the ACM* 37, no. 2 (February 1994): 21–5; Ashurst, 'First World Smartcards and Third World Pensioners'.

[107] 'R700000 for Armoured Cars', *The Citizen*, 16 September 1993.

[108] Staff Reporter, 'Tokyo Sexwale's Prints All over R10-Billion Tender', *The M&G Online*, 23 March 2012, http://mg.co.za/article/2012-03-23-sexwales-prints-all-over-r10bn-tender/.

[109] Rehana Rossouw, 'The Payment That Put Paid to Abe', *Weekly Mail*, 23 February 1996.

competing firms and flimsily documented press reporting. The cumulative result of these controversies – much more than the older accusations of official fraud – has been to significantly damage the legitimacy of the elected government, and the reputation of CPS in particular.[110]

## Comparisons and conclusions

In the decade since their first introduction in the middle of the 1990s the countries that have adopted direct cash transfers, either as temporary aid or as a permanent source of social benefit, have moved to the biometric registration and payment systems that were first developed in Zululand.[111] The causalities at work here are complex, but it is clear that outside of those societies with well-developed systems of civil registration the movement towards biometric citizenship has taken on what Hughes described as technological momentum – where the networks of human, physical and financial agents act together to move towards a common and discernible outcome.[112] The specific goals, timing and sequence of the arrangements differs significantly from country to country, but a basic model of centralised identity registration, private financial control and automated local cash transfers, all using fingerprint biometrics, has emerged in each case.

Consider Brazil's *Bolsa Família* (or Family Allowance) which is widely regarded as the international exemplar of successful redistributive poverty reform.[113] The first direct cash grants were introduced there by individual municipalities in 1995, and they were aimed at encouraging poor children to attend school (a goal that remains key to the Brazilian policy). In 1999 the Cardoso government introduced federal subsidies for food,

---

[110] Staff Reporter, 'Tokyo Sexwale's Prints All over R10-Billion Tender'; 'Sassa Judgment Illegal but Won't Be Set Aside', *The M&G Online*, 28 August 2012, http://mg.co.za/article/2012-08-28-sassa-ruling-illegal-but-wont-be-set-aside/; Thabiso Mochiko, 'Net1 Given Leave to Appeal Social Grants Tender Verdict', *Business Day Live*, 14 September 2012, www.bdlive.co.za/business/technology/2012/09/14/net1-given-leave-to-appeal-social-grants-tender-verdict; Mamello Masote, 'Absa Won't Quit Net1 Welfare Tender Battle', *Business Day Live*, 31 March 2013, www.bdlive.co.za/business/financial/2013/03/31/absa-won-t-quit-net1-welfare-tender-battle; Staff Reporter, 'Absa, AllPay Lose Court Bid over Tender', *Sunday World*, 1 April 2013, www.sundayworld.co.za/news/2013/04/01/absa-allpay-lose-court-bid-over-tender.

[111] For the global spread of these systems in the former colonial world, see Gelb and Clark, 'Identification for Development'; and Gelb and Decker, 'Cash at Your Fingertips'.

[112] Thomas Hughes, 'Technological Momentum', in *Does Technology Drive History? The Dilemma of Technological Determinism*, ed. Michael L. Smith and Leo Marx (Cambridge, MA: MIT Press, 1994), 101–13.

[113] 'Latin America: Gini Back in the Bottle', *The Economist*, 13 October 2012, www.economist.com/node/21564411.

cooking gas and school attendance to a limited number of the poorest families. The grants were initially arranged and monitored by municipal schools and teachers were the most important registrars. This supported very onerous forms of moralisation and encouraged local practices of denunciation for the undeserving or uncompliant.[114] From 2002, under the Lula government, the mix of grants began to evolve towards a single family allowance.[115]

The consolidation of the four existing cash grants into a single family allowance delivered by means of an electronic benefits card in 2003 coincided with the development of a single national database of beneficiaries, the Cadastro Unico. This database was significantly more complex than anything currently demanded by the South African government – indeed in its curiosity about the beneficiaries it resembles the failed panoptic ambitions of the Home Affairs project – but it did have the effect of moving the work and regulation of registration from the municipalities up to federal government.[116] Unlike South Africa, municipalities – and there are over 5,000 of them in Brazil – remain key to the workings of *Bolsa Família*, and some of the wealthiest supplement the grants with payments of their own. The Brazilian (and Mexican) grants are also very much smaller in real terms and as proportions of GDP – on both measures the South African grant is ten times as large as the others.[117]

But in other respects *Bolsa Família* has evolved to become very like the South African system. After 2003 the grants were paid and administered centrally by the Caixa Economica Federal (CEF), the second largest and government-owned bank; individuals were identified using a Social Identification Number, and the payments were transferred electronically to cards that accessed the bank's national network of automated teller machines.[118] In July 2012 CEF announced that it would be installing biometric authentication readers on its ATMs, for reasons that echo the

[114] A. De Janvry, F. Finan, E. Sadoulet, D. Nelson, K. Lindert, B. de la Brière and P. Lanjouw, 'Brazil's Bolsa Escola Program: The Role of Local Governance in Decentralized Implementation', Washington, DC: World Bank, Social Protection (SP), SP Discussion Paper no. 0542 (2005), http://siteresources.worldbank.org/SOCIALPROTECTION/Resources/SP-Discussion-papers/Safety-Nets-DP/0542.pdf.

[115] K. Lindert, A. Linder, J. Hobbs and B. de la Brière, 'The Nuts and Bolts of Brazil's Bolsa Família Program: Implementing Conditional Cash Transfers in a Decentralized Context', (Washington DC: The World Bank, 2006): 18.

[116] Bénédicte de la Brière and Kathy Lindert, *Reforming Brazil's Cadastro Único to Improve the Targeting of the Bolsa Família Program*, Social Protection Discussion Paper Series (Washington DC: Social Protection Unit, Human Development Network, The World Bank, June 2005), www.mef.gob.pe/contenidos/pol_econ/documentos/Bolsa_de_Familia_Brasil.pdf; De Janvry *et al.*, 'Brazil's Bolsa Escola Program', 38–9.

[117] Hanlon *et al.*, *Just Give Money to the Poor*, 155.

[118] Lindert *et al.*, 'The Nuts and Bolts of Brazil's Bolsa Família Program'.

explanations in South Africa. 'Many users in the program do not have bank accounts and use the ATM only once a month to get their stipend', the contractor explained, 'as such, they often forgot their passwords and bank managers were spending too much time resolving getting PINs renewed or changed.'[119]

To date there are no plans to introduce a central biometric population register in Brazil (where civil registration has long been a fraught and unreliable tool of elaborately developed institutions of local government),[120] but in almost all of the major former colonies plans for centralised biometric systems identical to the South African population register are well under way.[121] The purest of these projects is certainly the UID plan in India, where Nandan Nilekani is driving the development of a single centralised repository of biometric identity that is intended to work as the radically simplified central government lynchpin over an infrastructure of grants and credit that mimics the existing South African economy.[122] There is more than a little irony in the fact that systems of biometric registration are returning home a century after they were brought to South Africa from India.

---

[119] 'Brazilian Bank CAIXA Deploys Lumidigm Fingerprint Sensors in ATMs', 23 July 2012, www.lumidigm.com/brazilian-bank-caixa-deploys-lumidigm-fingerprint-sensors-in-atms/.
[120] Mara Loveman, 'Blinded Like a State: The Revolt Against Civil Registration in Nineteenth-Century Brazil', *Comparative Studies in Society and History* 49, no. 01 (2007): 5–39; Wood and Firmino, 'Empowerment or Repression?'
[121] Garcia and Moore, *The Cash Dividend.*
[122] For the details and implications see Gelb and Clark, 'Identification for Development'.

# Epilogue: empire and the mimetic fantasy

Over the last decade the newly created US Secretariat of Homeland Security has been a powerful advocate of global biometric identification. It was Michael Chertoff who occupied this post in the second term of the Bush administration. In early May, 2007, he addressed an audience of students at the Johns Hopkins School for Advanced International Studies (SAIS) on the subject of 'Addressing Transnational Threats in the 21st Century'. It is worth noting that SAIS is one of the key sites for the training of professional diplomats and that the students there can fairly be described as experts in international relations and government. Chertoff's speech, unlike many others on the subject of the War on Terror, was a statement of what the Bush administration believed was wise and practical. And it demonstrated that, long after the initial drama of the attacks of 9/11, global biometric registration remained key to US domestic security policy.

Chertoff spoke eloquently of the confounding effects of globalisation on the US government's efforts to identify and combat its current enemies; invoking the Cold War doctrine of defence-in-depth he argued that the most important policy goal was 'extending the protection of the perimeter'. Information, he argued, is the twenty-first century equivalent of the massive radar systems that guarded the borders of the continental United States during the Cold War, because it will allow 'us to isolate the individual who is a threat from the great mass of people coming in who are innocent'. In this struggle over the terrain of information the United States will exploit its technological ascendancy through the deployment of biometric identification systems like US-Visit, the programme that collects biometrics from visa applicants, which he claimed allowed for the matching of fingerprints at the points of entry against existing criminal and terrorist databases.

But the US plans for biometric registration extend well beyond immigration control to a global system of fingerprint gathering. 'We're moving to 10-print collection overseas and at our ports of entry, which will allow us one day in the very near future to check a visitor's or a potential

196

visitor's fingerprints against latent fingerprints that we collect in bat-
tlefields and safehouses all around the world.' Anticipating the obvious
question of whether such a system could ever be made to work Chertoff
explained that a vigilant INS agent at O'Hare Airport had recently
refused entry to a suspect visitor, sending him 'back to where he came
from' after recording his fingerprints. 'We did ultimately run across those
fingerprints again', he explained to the students, 'at least parts of the
fingerprints, because a couple years later we found them on the steering
wheel of a suicide truck bomb that had been detonated in Iraq.'[1]

There are some odd things about this speech. Chertoff was massively
overstating the speed and power of biometric databases. Rapid integrated
searching of large databases was not possible at that time, although it
is now. As late as March 2006 the National Institute of Standards and
Technology (NIST), under pressure from the US Congress, was bat-
tling to define a single technical standard that would allow the different
commercial systems owned by the FBI and the Department of State to
interact accurately and efficiently.[2] Just months before Chertoff's speech
NIST noted that even the most carefully compiled ten-print systems
were incapable of fully automated matching, requiring the intervention
of a human fingerprint expert. Heavy federal government investment in
interoperability over the next five years did make it possible to compare
carefully taken, ten-print records between the FBI and immigration and
military systems.[3] But for single, or latent, fingerprint matching – the
kind that might be left on a bomb – almost all of the important work
has, still, to be done by human experts with obvious devastating effects
on the possibility of using biometric identification to process millions
of travellers against a collection of latent prints gathered from 'battle-
fields and safehouses'.[4] Chertoff, like every biometric enthusiast before
him, was also deliberately blurring the boundary between the statistically
clear identification by ten rolled fingerprints and the murky and treacherous waters of latent print similarity. Here identity can only be proven
by a court-sanctioned expert, and there is currently no scientific case

---

[1] Michael Chertoff, 'Remarks by Secretary Michael Chertoff to the Johns Hopkins
University Paul H. Nitze School of Advanced International Studies', *Department of
Homeland Security*, 3 May 2007, www.dhs.gov/xnews/speeches/sp_1178288606838.
shtm.
[2] National Institute of Standards and Technology, 'Minutiae Interoperability Exchange
Test 2004', 21 March 2006, http://fingerprint.nist.gov/minex04/.
[3] Subcommittee on Biometrics, *The National Biometrics Challenge* (Washington, DC:
National Science and Technology Council, 15 September 2011), 7–8, www.biometrics.
gov/NSTC/Publications.aspx.
[4] V.N. Dvornychenko and Michael D. Garris, *Summary of NIST Latent Fingerprint Testing
Workshop*, November 2006.

for unique identification by latent prints. The nasty truth – in a world of massively expanding fingerprint databases – is that the likelihood of false matches is proportional to the size of print populations and the level of political pressure on the examiners.[5]

Nor is this simply a matter of exaggeration; Chertoff was making claims for biometric registration that exist only in the domain of magic. He was invoking the mimetic power that has often captivated the advocates (and the subjects) of compulsory fingerprint registration. Similarity is one part of what makes this compelling; contact is another. Simon Cole has shown that the power of latent fingerprint identification in the courts 'lies in the seemingly magical ability to cause these stereoscopic images to merge in the jury's eyes into one'.[6] The same desire to close the gap between the fingerprint and the suspect clearly motivated Chertoff's account. He presented latent fingerprint matching as an infallible tool of global surveillance, blithely ignoring the similarity between his anonymous example and the abundantly documented Mayfield fiasco. In this case, Brandon Mayfield, a Muslim lawyer in Seattle, was wrongly arrested and charged with terrorism in Seattle in 2004 on the basis of a latent fingerprint found in Madrid. Mayfield's fingerprints were only in the FBI database because he had served eight, honourable, years in the US military.[7] More than anything his case demonstrated the real – although statistically unlikely – danger to innocent citizens of large-scale latent fingerprint searches. Yet this capricious danger was inverted in Chertoff's explanation, which makes what can only be described as fantastic claims for the certainty of latent fingerprint matches.

The magical qualities that Chertoff attributes to biometrics extend to several other areas: like radar, they will act as a hemispheric shield; they will give the US government the power to reach out, beyond the continental perimeter, into the safe-houses of its enemies; and, most importantly, to seize them by their likeness. This, as Taussig observed some time ago, is what makes mimesis the essence of sympathetic magic.[8] Francis

---

[5] David H. Kaye, 'Questioning a Courtroom Proof of the Uniqueness of Fingerprints', *International Statistical Review* 71, no. 3 (1 December 2003): 521–33; Simon A. Cole, 'Is Fingerprint Identification Valid? Rhetorics of Reliability in Fingerprint Proponents' Discourse', *Law & Policy* 28, no. 1 (January 2006): 109–35.

[6] Simon A. Cole, 'Witnessing Identification: Latent Fingerprinting Evidence and Expert Knowledge', *Social Studies of Science* 28, no. 5/6, Special Issue on Contested Identities: Science, Law and Forensic Practice (1998): 687–712; for a discussion of the fallibillity of LFPEs see Simon A. Cole, *Suspect Identities: A History of Fingerprinting and Criminal Identification* (Cambridge, MA: Harvard University Press, 2001), 281–3.

[7] Sarah Kershaw, 'Spain and US at Odds on Mistaken Terror Arrest', *New York Times*, 5 June 2004.

[8] Michael Taussig, *Mimesis and Alterity: A Particular History of the Senses* (New York: Routledge, 1993), 47, 221–3. Taussig acknowledges Pamela Sankar as the source of his insightful discussion of fingerprinting as mimesis.

Galton, writing from the epicentre of the imperial ethnographic project, was very aware of fingerprinting's mimetic power, simultaneously stressing and dismissing the 'abundant instances of the belief that personal contact communicates some mysterious essence from the thing touched to the person who touches it and vice versa'.[9] Galton's invention of fingerprinting was both a product and tool of the late nineteenth-century demands of imperial government, and Taussig was correct to suggest that 'subterranean notions' of the '"magic" of copy and contact' remained powerful elements in the politics of fingerprinting, motivating advocates to expand its powers and constraining those subject to it far beyond its actual reach.[10]

The current US Department of Homeland Security and its policy of global biometric surveillance rests heavily on an association with Britain, and the former empire. An alliance between the National Security Agency in the United States and British signals intelligence (and their equivalents in the former white colonies) has long provided the mechanisms for an enormous data-gathering project of communications across international borders.[11] There are some surprising and important political links here with the imperial project that produced the South African state. The US National Security Council, which as Hogan shows nurtured the frenzy over the global communist danger and the explosion of federal military spending after 1950, was modelled on the older (and constitutionally unprecedented) British Imperial War Cabinet. That extra-parliamentary committee was long the primary political goal of the architect of modern South Africa, Lord Alfred Milner. In the decades after the First World War, Milner's acolytes, relying on the resources of the South African gold magnates and relationships carefully nurtured by the same Lionel Curtis who features so prominently in this book, assisted by the Rhodes Trust and its troop of Scholars, together fostered the institutional and ideological basis of the current Anglo-American alliance.[12]

[9] Francis Galton, *Finger Prints* (London and New York: Macmillan and Co., 1892), 38–40.
[10] Taussig, *Mimesis and Alterity*, 221–3.
[11] See James Bamford, *The Puzzle Palace: A Report on America's Most Secret Agency* (London: Penguin, 1983), 309–37; David Lyon, *Surveillance after September 11* (Malden, MA: Polity Press in association with Blackwell Pub. Inc., 2003), 117; as Bamford shows, this massive international surveillance project proved singularly blind, leading to the attacks in 2001, which were largely arranged within sight of the NSA's massive headquarters in Maryland. James Bamford, *Body of Secrets: How America's NSA and Britain's GCHQ Eavesdrop on the World* (London: Arrow, 2002), 614–51.
[12] Michael J. Hogan, *A Cross of Iron: Harry S Truman and the Origins of the National Security State* (Cambridge University Press, 1998), 33, 68, 195; Franklyn A. Johnson, 'The British Committee of Imperial Defence: Prototype of U.S. Security Organization', *The Journal of Politics* 23, no. 2 (1961): 231–61; on the relationship between the NSC and the new Homeland Security Council, see William W. Newmann, 'Reorganizing

Chertoff acknowledged the ongoing significance of this close and cooperative alliance with Britain in the new global conflict, in deliberate contrast with 'those in Europe who feel that this principle of sharing ought not to be extended across the ocean'. Invoking the coordination between the two governments during the August 2006 panic over the possible use of liquids to attack commercial aircraft as a 'model of how two countries working together in partnership and trust can share information, bring down and disrupt a plot', Chertoff made repeated references to British support for his department's work in the Global War on Terror. His speech concluded by citing Peter Clark, the 'head of counter-terrorism for Scotland Yard', as an authority on the unprecedented danger posed by al Qaeda. Using Clark's authority he reminded his audience that 'this is a global threat of a kind not seen before'.

More recent developments in the technology of biometric surveillance emphasise these close connections between the old empire and the new world order. Early in 2008, after years of disagreement with the European Union over the content and form of personal data-sharing, the FBI proposed plans for a 'Server in the Sky' to share biometric data between the current allies in the War on Terror, the so-called Anglophone members of the British Commonwealth: Australia, Britain, Canada and New Zealand. This system would allow the IAFIS database owned and controlled by the FBI to interact with IDENT1, the biometric repository controlled by the British National Policing Improvement Agency. One of the reasons that this integration was possible was that both database infrastructures were being supplied by the same company. Northrop Grumman, one of the major suppliers in the field of modern biometrics, was contracted to supply the British police system and the new connections between the FBI and US immigration databases.[13] The

for National Security and Homeland Security', *Public Administration Review* 62, no. s1 (2002): 126–37; Lionel Curtis was the key figure in the fashioning of this embrace, see D. Lavin, *From Empire to International Commonwealth: A Biography of Lionel Curtis* (New York: Oxford University Press, 1995), 161–77. And the links between this Anglo-American world order and South African politics are much more direct than many people realise. One of the outstanding architects of the twentieth-century global order, as Mitchell and Mazower have each separately shown in important recent studies of very different global institutions, was Jan Smuts, founder and builder of the South African state. Mark Mazower, *No Enchanted Palace: The End of Empire and the Ideological Origins of the United Nations*, Lawrence Stone Lectures (Princeton University Press, 2009); Timothy Mitchell, *Carbon Democracy: Political Power in the Age of Oil* (London: Verso, 2011).

13  Richard Koman, 'Server in the Sky: FBI International Biometric Db Planned', News, *ZDNet*, 14 January 2008, http://government.zdnet.com/?p=3605; Owen Bowcott, 'FBI Wants Instant Access to British Identity Data', *Guardian*, 15 January 2008, www.guardian.co.uk/uk/2008/jan/15/world.ukcrime; 'Britain's Police Balk at Plug-in

Commonwealth countries named in the FBI's proposal promptly disa-vowed the FBI's data-sharing arrangements – under pressure from the same popular worries over privacy that limit the expansion of biometric surveillance on the US mainland – but, as the Snowden affair amply demonstrates, the infrastructural connections between US cross-border surveillance and the British empire run deep.[14]

Under the Obama administration some of the crudest technologi-cal enthusiasm for biometrics has been dampened by a much broader interest in 'smart power', which professes to respect the rule of law, and the social causes of radicalisation and looks to use USAID (and social media) as instruments of anti-terrorist policy.[15] Yet, even at the heart of this socially oriented foreign policy, biometric screening remains the most important element of US border security. The Obama administra-tion remains vigorously committed to solving the technical difficulties that prevent real-time biometric sensors from interacting properly with the largest federal databases. In some respects this effort actually exceeds the goals of the Bush administration, pushing the plans of the federal government close to the dystopian themes of science fiction. Perhaps the most fraught development in the current US biometrics programme is the effort to develop 'portable rapid DNA machines' that, according to the NTSC's Subcommittee on Biometrics (an organisation that tends, unusually, to pessimistic assessments) is 'poised to provide a new tool for rapid identification outside of the forensic laboratory'.[16] It is important to notice the continuing military and national security emphasis of the US interest in biometrics – which sees soldiers routinely gathering fin-gerprints from bomb fragments, from civilians in Iraq and Afghanistan and diplomats collecting them from police officials and embassies in Egypt – as it is, also, obvious that the US project now finds itself in the territories formerly occupied by Britain.[17]

to FBI Database', *Washington Times*, 16 January 2008, www.washingtontimes.com/news/2008/jan/16/britains-police32balk-at-plug-in32to-fbi-database/; Lewis Page, 'UK.gov Says No Plans for FBI DNA Database Hookup', *The Register*, 17 January 2008, www.theregister.co.uk/2008/01/17/fbi_uk_dna_database_plans_followup/; Mark Russell, 'FBI Invites Australia to Join World Crime Database', *The Age*, 20 January 2008, www.theage.com.au/news/national/fbi-invites-australia-to-join-world-crime-database/2008/01/19/1200620280804.html; Rebecca Palmer, 'NZ Police May Join FBI Network', *Stuff.co.nz*, 15 September 2008, www.stuff.co.nz/4357650a11.html.

[14] Ewen MacAskill, Julian Borger, Nick Hopkins, Nick Davies and James Ball, 'GCHQ Taps Fibre-Optic Cables for Secret Access to World's Communications', *Guardian*, 21 June 2013, www.guardian.co.uk/uk/2013/jun/21/gchq-cables-secret-world-communications-nsa.

[15] Hillary Clinton, 'Smart Power Approach to Counterterrorism' (John Jay School of Criminal Justice, 9 September 2011), www.state.gov/secretary/rm/2011/09/172034.htm.

[16] Subcommittee on Biometrics, *The National Biometrics Challenge*, 14.

[17] *Ibid.*; 'US Diplomats Spied on UN Leadership', *Guardian*, 28 November 2010, www.guardian.co.uk/world/2010/nov/28/us-embassy-cables-spying-un; 'WikiLeaks Raises

There are many current and past examples of biometric registration targeted at domestic populations in the United States and Europe. These have included surprisingly successful campaigns in the United States for voluntary fingerprint registration in the 1930s, the compulsory registration of government employees, of members of the military, and even the licensing of some professionals.[18] Recently the most widely distributed systems of biometric identification in these countries have been applied to passports, typically involving the use of facial images, sometimes of fingerprints.[19] But it is important to notice that these are hobbled biometrics, carefully restricted to simple one-to-one matches of existing documents of identity. As Gandhi might explain, they answer the question, 'Is the bearer of this document the Joan Smith who originally applied for it?' And they are very different from the one-to-many, omniscient and omnipotent systems being designed for the criminal justice system, for immigration and for national security. Those systems answer the question, 'Who is this?' And they have been specifically developed to strip Joan of any agency in answering the question.

It is also true that the US Federal Bureau of Investigation has been the primary custodian of fingerprinting through most of the last century, and of automated fingerprint identification in the last generation.[20] And the effects on US society and politics have been important, bearing interesting comparisons with the machine of incarceration that lay at the heart of the Apartheid state. The automation of fingerprint registration over the last thirty years has certainly been an important part of the startling expansion of compulsory and indefinite imprisonment in the United States. Nor is it incidental that a very large proportion of those targeted are black.[21]

Much the same can be said about social welfare. In both the United States and Britain the use of fingerprinting for the identification and control of welfare recipients has been long contemplated by reformers and administrators.[22] But it was only very recently, with the efficiencies that

Specter Of Biometric Data', *All Things Considered* (NPR, 30 November 2010), www.npr.org/2010/11/30/131704360/wikileaks-raises-specter-of-biometric-data.

[18] Cole, *Suspect Identities*; P. Sankar, 'State Power and Record-Keeping: The History of Individualized Surveillance in the United States, 1790–1935' (University of Pennsylvania, 1992).

[19] Louise Amoore, 'Biometric Borders: Governing Mobilities in the War on Terror', *Political Geography* 25, no. 3 (March 2006): 336–51; Mark Maguire, 'The Birth of Biometric Security', *Anthropology Today* 25, no. 2 (2009): 9–14.

[20] Cole, *Suspect Identities*, 248–58.

[21] Simon Cole, 'Digits – Automated Fingerprinting and New Biometric Technologies' (Unpublished paper, n.d.), 21; D. Garland, *The Culture of Control: Crime and Social Order in Contemporary Society* (New York: Oxford University Press, 2001).

[22] Edward Higgs, *Identifying the English: A History of Personal Identification, 1500 to the Present* (London and New York: Continuum, 2011), 145–9; June Purcell Guild, 'Transients in

computerised identification brought to fingerprinting, that large-scale social welfare schemes in these countries have been organised biometrically. These technologies were brought to the United States from South Africa. It is also important to notice that the application of biometrics to social welfare has coincided with a fierce public assault on the status of welfare recipients as citizens.[23]

There is a common pattern here. The largest centralised systems in the criminal justice, social welfare and immigration control systems in the United States and Europe have been designed to target individuals and populations that have been significantly stripped of the rights and statuses of citizens. When important administrators have proposed compulsory biometric registration for the universal social security entitlements of all citizens in the United States (and the UK) these schemes have conspicuously fallen on deaf ears.[24] In this, biometric registration in the northern hemisphere is very different from the massive and centralised national population registers of biometric identification that were first developed in South Africa, and which are now being developed throughout what used to be called the Third World. These new instruments of biometric citizenship target all citizens, and they are designed to do the work of civil registration, and, especially, to regulate identification in financial transactions.

Biometric identity registration in Europe, the United States and Australia has retreated in the face of widespread popular protest. This raises the question of why the same systems, busily under way in many of the former colonies, have been so easily defeated in the wealthy liberal democracies.[25] Several common features emerge. The first significant

a New Guise', *Social Forces* 17, no. 3 (1939): 366–72; Myron Falk, 'Fingerprints: Black Marks against the Migrant', *Social Forces* 19, no. 1 (1 October 1940): 52–6.

[23] Anna Marie Smith, *Welfare Reform and Sexual Regulation* (Cambridge University Press, 2007); Shoshana Magnet, 'Bio-Benefits: Technologies of Criminalization, Biometrics and the Welfare System', in *Surveillance and Social Problems*, 2008, www.magnetopia. org/biometrics%20and%20welfare.doc; Harry Murray, 'Deniable Degradation: The Finger-Imaging of Welfare Recipients', *Sociological Forum* 15, no. 1 (1 March 2000): 39–63.

[24] Doris Meissner and James Ziglar, 'The Winning Card', *New York Times*, 16 April 2007, www.nytimes.com/2007/04/16/opinion/16meissner.html?scp=103&sq=biometric&st=cse.

[25] Pierre Piazza and Laurent Laniel, 'The INES Biometric Card and the Politics of National Identity Assignment in France', in *Playing the Identity Card: Surveillance, Security and Identification in Global Perspective*, ed. David Lyon and Colin Bennett (London and New York: Routledge, 2008), 93–111; Dean Wilson, 'The Politics of Australia's "Access Card"', in *Playing the Identity Card: Surveillance, Security and Identification in Global Perspective*, ed. David Lyon and Colin Bennett (London and New York: Routledge, 2008), 180–97; Kelly Gates, 'The United States Real ID Act and the Securitization of Identity', in *Playing the Identity Card: Surveillance, Security and Identification in Global Perspective*, ed. David Lyon and Colin Bennett (London and New York: Routledge, 2008), 218–232;

difference is that the main organisers of the resistance to biometrics in the West have been engineers and scientists who have a clear understanding of the likelihood, and implications, of system failures.[26] These figures have typically worked in close alliance with individuals in the media who write for a public with a well-honed contempt for bureaucratic hubris, and a lively interest in the sordid – and often very amusing – details of official administrative bungling. And at the core of this public scepticism of the state's will to know its subjects lies the disorganised, contradictory but (as Solove shows) nonetheless very powerful political theory of the right to privacy.[27] This is a field of government and law that has grown dramatically in authority and scope over the last three decades in the northern hemisphere.[28] The same cannot be said of the South, where the right to privacy is routinely presented as an unsustainable casualty of the project of survival.[29] Of course, that no similar body of law or regulation exists in the former colonies has much to do with the debased place of native privacy under imperial government.

The combination of scientific criticisms with an entrenched two-party democracy is also important. On both sides of the Atlantic the opponents of biometric registration have drawn allies from both the left and the right, making it very difficult for social engineers to sustain the political power required to drive through large-scale registration projects. These democratic limits work in part because of the mobilising fear of an assault on established privacy rights, in part because both the right and the left fear the surveillance implications of biometrics, and in part because compulsory fingerprinting cannot shed its particular sentimental and ideological history. 'All words', Bakhtin wrote, 'have the "taste" of a profession, a genre, a tendency, a party, a particular person, a generation, an age group, the day and hour.'[30] And fingerprinting, as Karl Pearson lamented in 1930, has long been 'tainted in the popular mind by

'Last Rites for ID Cards Read by Johnson', *Independent*, 1 July 2009, www.independent. co.uk/news/uk/home-news/last-rites-for-id-cards-read-by-johnson-1726187.html.

[26] Ross Anderson, Ian Brown, Terri Dowty, Philip Inglesant, William Heath and Angela Sasse, *Database State* (York: Joseph Rowntree Reform Trust, 2009); Whither Biometrics Committee, National Research Council, *Biometric Recognition: Challenges and Opportunities*, ed. Joseph N. Pato and Lynette I. Millett (Washington, DC: The National Academies Press, 2010).

[27] Daniel J. Solove, '"I've Got Nothing to Hide" and Other Misunderstandings of Privacy', *San Diego Law Review* 44 (2007): 745–72.

[28] Colin J. Bennett and Charles D. Raab, *The Governance of Privacy: Policy Instruments in Global Perspective* (Aldershot: Ashgate, 2003); Fred H. Cate, *Privacy in the Information Age* (Washington, DC: Brookings Institution Press, 1997).

[29] Claude Ake, 'The African Context of Human Rights', *Africa Today* 34, no. 1/2 (1 March 1987): 5–12.

[30] M.M. Bakhtin, *Dialogic Imagination: Four Essays* (Austin: Texas University Press, 1981), 293.

a criminal atmosphere'.[31] Imperial subjection has been another key part of the taste of fingerprint identification.

In his recent history of identification in England, Eddy Higgs shows how written forms of identification very gradually became markers of respectability as literacy spread from the fourteenth century.[32] By the nineteenth century written forms of identification and written contracts had become key signs of English respectability. These forms of identification built on the long history and practically universal networks of written civil-registration in the parishes.[33] 'Identification through the body', Higgs writes, 'was associated with the nonrespectable, the deviant, the foreign and the alien'.[34] Yet in the first half of the nineteenth century identification in the empire relied heavily on the marking of the body itself. Branding, mutilation and tattooing were important weapons in the arsenal of policing and imprisonment in India well into the 1840s, and they were remembered nostalgically for decades afterwards.[35] And it was from these techniques of marking the body, themselves brutally mimetic, that the early forms of fingerprinting developed in India.[36] As Higgs points out, the rejection of fingerprinting in England after 1920 stemmed in part from the fact that both officials and the public remembered the history, and abjection, of fingerprinting in South Africa.[37] It was this political reputation and context that determined the norms of official documentation in the North and the South. Citizenship mediated by writing remains the norm for most people in the liberal democracies, while a mimetic state has begun to develop in the old colonies.

## Imperial progressivism

It is easy to view the recent expansion of the American state's global surveillance ambitions as an unprecedented consequence of the attacks on

---

[31] Karl Pearson, *The Life, Letters and Labours of Francis Galton: Correlation, Personal Identification and Eugenics*, vol. 3A (Cambridge University Press, 1930), 159.

[32] Higgs, *Identifying the English*, 60–72.

[33] Simon Szreter, 'Registration of Identities in Early Modern English Parishes and Amongst the English Overseas', in *Registration and Recognition: Documenting the Person in World History*, ed. Keith Breckenridge and Simon Szreter, Proceedings of the British Academy 182 (Oxford University Press, 2012), 67–92.

[34] Higgs, *Identifying the English*, 77.

[35] Radhika Singha, 'Settle, Mobilize, Verify: Identification Practices in Colonial India', *Studies in History* 16, no. 2 (2000): 151–98; Clare Anderson, *Legible Bodies: Race, Criminality, and Colonialism in South Asia* (Oxford and New York: Berg, 2004), 42.

[36] Chandak Sengoopta, *Imprint of the Raj: How Fingerprinting Was Born in Colonial India* (London: Macmillan, 2003), 73–5.

[37] Edward Higgs, 'Fingerprints and Citizenship: The British State and the Identification of Pensioners in the Interwar Period', *History Workshop Journal* 69 (2010): 62.

the Twin Towers. This idea – that the state was jolted from an easy-going cosmopolitan rest by the New York attacks – is one of the organising claims of the so-called Global War on Terror. Yet in accepting this view we can lose sight of some intriguingly enduring features of American government which are pertinent to this history. The idea that biometric registration can work to protect the mainland from foreign enemies is one of these older practices, central to the Protestant, middle-class nativism that Hofstadter complained about fifty years ago.[38] Perhaps the best example of this obsession dates from the middle of the last century. It was under the terms of the Smith Act of 1940 – the infamous law that equipped J. Edgar Hoover and Joseph McCarthy with the legal tools for the national anti-communist witch-hunt of the 1950s – that all foreigners entering the United States were first required to register their fingerprints.[39] This stipulation applied – much to the horror of British Members of Parliament – to all tourists, including those from the United Kingdom. It was only in October 1957 that foreigners staying in the United States for less than a year were forgiven the requirement to provide their fingerprints.[40]

Fingerprint registration has long been an important part of the plans of the intellectual movement historians now call Atlantic progressivism. Dating from the last decade of the nineteenth century to the last years of the 1920s, progressivism has been the subject of an enormous historiography (especially in the United States) that was elegantly synthesised by Daniel Rodgers in a study of the intellectual debates that shaped Atlantic societies in this period.[41] Rodgers shows that progressivism 'was English before it was American, born in the heated municipal politics of 1890s

---

[38] Richard Hofstadter, *The Age of Reform: From Bryan to FDR* (New York: Alfred A Knopf, 1956).

[39] Ellen Schrecker, *Many Are the Crimes: McCarthyism in America* (Boston: Little Brown and Company, 1998), 97; Richard Gid Powers, *Secrecy and Power: The Life of J Edgar Hoover* (London: Hutchinson & Co, 1987), 238.

[40] 'United States Visa Regulations' (*Hansard*, UK, 23 March 1954), http://hansard.millbanksystems.com/lords/1954/mar/23/united-states-visa-regulations#S5LV0186P0_19540323_HOL_13; 'Visa Formalities (Finger Prints)' (*Hansard*, UK, 8 March 1954), http://hansard.millbanksystems.com/commons/1954/mar/08/visa-formalities-finger-prints#S5CV0524P0_19540308_HOC_94; van Schalkwyk, Minister, South Africa House to Secretary for External Affairs, 'Parliamentary Question: Fingerprinting of British Subjects Applying for Visas', 4 December 1957, BNS 1/1/328, 42/74 Fingerprints and Photographs on Permits Etcetera Issued to Indians. General Questions. Part 2. 1928–1957, SAB.

[41] D.T. Rodgers, *Atlantic Crossings: Social Politics in a Progressive Age* (Cambridge, MA: Harvard University Press, 2000); before Rodgers the foundational studies of progressivism were: Hofstadter, *Age of Reform*; Robert H. Wiebe, *The Search for Order, 1877–1920* (New York: Hill and Wang, 1967); C. Vann Woodward, *The Strange Career of Jim Crow* (New York: Oxford University Press, 1966).

London before crossing to the United States in the first decade of the new century' and that it set the foundations of the twentieth-century state in many countries on the Atlantic basin.

Much more than Foucault's very general (and historically obscure) account of statistically driven governmentality, physiologically motivated biopower, and even more than Scott's authoritarian high modernism, early twentieth-century progressivism set the foundations and ambitions of the modern state.[42] It was Beatrice Webb – perhaps the most prolific of the progressives – who coined the term we now associate with state-planning run amok. Commenting in 1918 on the likely reception of the new book she and Sidney Webb had completed on *A Constitution for the Socialist Commonwealth of Great Britain* she observed that 'we shall offend all sides and sections with some of our proposals, but someone must begin to think things out, and our task in life is to be pioneers in social engineering'.[43] It was certainly progressivism that motivated the real and very energetic engineers of the early twentieth century, the system build-ers that fashioned the new corporations and municipal governments.[44] And it was these figures, especially the American mining engineers like Herbert Hoover, that helped carry the doctrines of progressive reform out to the outposts of the British empire.

Like post-modernity, progressivism defined an epoch and a glo-bal intellectual movement; there was nothing approaching a consensus about its political aims at the time, nor has one developed subsequently. It was common to find key progressives – like Woodrow Wilson and Theodore Roosevelt, or Joseph Chamberlain and Beatrice Webb – on opposite sides of the political divide. The movement was, as Rodgers

---

[42] Peter Burchell, Colin Gordon and Peter Miller, eds, *The Foucault Effect: Studies in Governmentality with Two Lectures by and an Interview with Michel Foucault* (London: Harvester Wheatsheaf, 1991); for detailed studies of the effects of progressivism on the twentieth-century state, see Theda Skocpol, *Protecting Soldiers and Mothers: The Political Origins of Social Policy in the United States* (Cambridge, MA: Harvard University Press, 1992); John W. Cell, *The Highest Stage of White Supremacy: The Origins of Segregation in South Africa and the American South* (Cambridge University Press, 1982); Simon Szreter, *Fertility, Class and Gender in Britain, 1860–1940* (Cambridge University Press, 2002); G.S. Jones, *Outcast London: A Study in the Relationship between the Classes in Victorian Society* (New York: Pantheon, 1971).

[43] Beatrice Potter Webb, *'The Power to Alter Things,' 1905–1924*, ed. Norman Ian MacKenzie and Jeanne MacKenzie (Cambridge, MA: Belknap Press of Harvard University Press, 1984), 357.

[44] David F. Noble, *America By Design: Science, Technology and the Rise of Corporate Capitalism* (Oxford University Press, 1977); Edwin T. Layton, *The Revolt of the Engineers; Social Responsibility and the American Engineering Profession* (Cleveland: Press of Case Western Reserve University, 1971); on system builders, Paul N. Edwards, *A Vast Machine: Computer Models, Climate Data, and the Politics of Global Warming* (Cambridge, MA: MIT Press, 2010), 9–12.

stresses, less concerned with interests than it was with practical problems of social reform. And it was driven in large part by the apparently irresistible flow of legislative reforms from one country to the next. Yet a common set of preoccupations emerged unmistakably. Progressives worried about the moral effects of the new, enormous and squalid cities and the demoralising forms of work produced by the factories of Toynbee's Industrial Revolution. And they feared the corrupting effects of working-class patrimonialism and monopoly power on democracy. The were usually impatient and dismissive of the virtues of individualism, of liberal political economy and, most powerfully, of the reactionary interventions of the courts and the law. In contrast with the utilitarians, the totems of the movement were expert science – especially empirical social science and statistics – an obsessive concern with efficiency, and hard work. And, very often, they used a romantically framed concern for white racial health to justify the key elements of the modern welfare state: reform of the Dickensian poor law, limiting work hours, unemployment insurance, pensions for the aged, support for poor mothers, a national minimum wage, and, eventually, socialised public health.[45]

If many of the positive achievements of the modern welfare state can be traced to the progressives, so too can the dark side, and, especially, the twentieth-century enthusiasm for segregation. On both shores of the Atlantic middle-class reformers had begun their efforts with Settlement Houses carefully placed to allow the young activists to observe and correct the behaviour of the poor. This interest in moralising reform developed quite rapidly into an interest in isolating and sanitising the population they began to consider as inherently unfit.[46] Typically these were defined racially as immigrant populations. Strategies for identifying and segregating the undeserving poor were proposed on both sides of the Atlantic that stressed coercive and centralised systems of registration, compulsory labour exchanges, and labour colonies. Many of the most important figures of the movement also adopted the biological obsessions of Galton's eugenics but, importantly, not all of them. Yet, like Galton, the progressives ultimately failed in the effort to impose segregationist institutions on their own citizens, coming, instead, to

---

[45] Rodgers, *Atlantic Crossings*; Beatrice Potter Webb, *'Glitter Around and Darkness Within,' 1873–1892*, ed. Jeanne MacKenzie and Norman Ian MacKenzie (Cambridge, MA: Belknap Press of Harvard University Press, 1982); Webb, *The Power to Alter Things*; G. R. Searle, *The Quest for National Efficiency: A Study in British Politics and Political Thought, 1899–1914* (Oxford: Basil Blackwell, 1971); M.E. McGerr, *A Fierce Discontent: The Rise and Fall of the Progressive Movement in America, 1870–1920* (New York: Free Press, 2003); Jose Harris, *William Beveridge: A Biography*, rev. edn (Oxford: Clarendon Press, 1997).

[46] Standish Meacham, *Toynbee Hall and Social Reform, 1880–1914: The Search for Community* (New Haven: Yale University Press, 1987), 94–114.

rely on the compulsory biometric registration of alleged criminals and immigrants.[47]

In Britain the key figures in this effort were Galton and Sir Edward Henry, Commissioner of both the Bengal and the London Metropolitan Police. Both played important parts in the South African story, and I have already said a lot about them. Here I will confine myself to the other international advocates of biometric identification. In the United States a massive fingerprinting effort was famously the primary responsibility of J. Edgar Hoover, Director of the Federal Bureau of Investigation for the half-century after 1924. Hoover's appointment was part of a much broader progressive transformation of the US federal government fostered by the mining engineer Herbert Hoover.[48] A carefully crafted monopoly over the records of fingerprinting was the instrument of J. Edgar Hoover's transformation of the FBI. It served as the basis of a new set of national standards and statistical returns to position the FBI at the centre of the dispersed and, until that time, very disorganised system of policing that emerged from the cities and states.[49] In his seminal biography of the most influential federal government employee of the twentieth century, Richard Powers demonstrated the centrality of progressivism in the making of the FBI and its director: Hoover, whose 'education, brains, memory, even his game of golf were all seen as marks of the progressive business manager, who shared with the progressive movement a fascination with problems of organisation, efficiency and control'.[50] Indeed, as a freemason with an unshakable confidence that science, efficiency and rigidly maintained racial segregation could stem the tides of immigrant and urban corruption, J. Edgar Hoover's resemblance to the archetypal progressives of Richard Hofstadter's *Age of Reform* was, as Powers observed, 'uncanny'.[51] Hoover's life-work has had obvious effects on the form of the US state, both domestically and internationally. The

---

[47] Michael Freeden, 'Eugenics and Progressive Thought: A Study in Ideological Affinity', *Historical Journal* 22, no. 3 (1979): 645–71; D.A. MacKenzie, *Statistics in Britain, 1865– 1930: The Social Construction of Scientific Knowledge* (Edinburgh: Edinburgh University Press, 1981); G.R. Searle, *Eugenics and Politics in Britain, 1900–1914* (Leyden: Noordhoff International Publishing, 1976); D.J. Kevles, *In the Name of Eugenics: Genetics and the Uses of Human Heredity* (Cambridge, MA: Harvard University Press, 1995); Cole, *Suspect Identities*, 155–9; on the resistance to eugenics see Szreter, *Fertility, Class and Gender in Britain*, 240–66; and on the failure of biometric proposals in Britain, Higgs, *Identifying the English*, 145–50.

[48] James D. Calder, *The Origins and Development of Federal Crime Control Policy: Herbert Hoover's Initiatives* (Westport: Praeger, 1993); Joan Hoff Wilson, *Herbert Hoover, Forgotten Progressive* (Boston: Little, Brown, 1975); Noble, *America By Design*.

[49] Powers, *Secrecy and Power*, 155; Sankar, 'State Power and Record-Keeping', 290–300.

[50] Powers, *Secrecy and Power*, 145; Hofstadter, *Age of Reform*, 144.

[51] Powers, *Secrecy and Power*, 519, n 3.

current American enthusiasm for fingerprinting foreigners, far from being a product of the events of 2001, is part of a continuous effort to control immigration that has institutional and ideological roots in the nativist anxieties of the progressive era. But it is important to note that, like Britain, the fingerprinting effort has mostly faced outwards, targeting legal and illegal immigrants, and focusing domestically on marginal populations of criminals and welfare recipients.

Argentina was another Atlantic society shaped by a determined and enduring interest in the use of fingerprinting to confront the disorder of massive immigration. Like South Africa, Argentina has served as a national incubator of biometric government, and the systems of fingerprinting adopted in many countries in Europe and throughout Latin America can be traced directly to developments there in the 1890s.[52] Like the United States, Britain and South Africa, a racially-inflected progressivism provided the intellectual context and motivation for the adoption of new forms of identity registration and surveillance. Yet in one distinguishing respect Argentina is very different from South Africa, and like the other societies on the Atlantic basin: the precocious efforts of the advocates of compulsory biometric registration faced determined opposition from the courts, and, more importantly, the legal tradition of assessing claims to citizenship based on reputational negotiations of respectability that were judged communally.[53]

In the last quarter of the nineteenth century Argentina experienced mass immigration that tested the basic principle of freedom of movement that, as Herzog shows, had been integral to Spanish law for 500 years.[54] Like Johannesburg, by 1914 Buenos Aires had become a city of foreigners.[55] To control what they saw as a crisis of crime and immigration native Argentine reformers turned to the techniques of involuntary and centralised identification that had been developed in France in the previous century. It was Juan Vucetich – a criminologist of Croatian

[52] J. Edgar Hoover, 'Criminal Identification', *The American Journal of Police Science* 2, no. 1 (1931): 16; L.R. Almandos, 'Identification in the Argentine Republic', *Journal of Criminal Law and Criminology (1931–1951)* 24, no. 6 (1934): 1098–101; Eduardo A. Zimmermann, 'Racial Ideas and Social Reform: Argentina, 1890–1916', *The Hispanic American Historical Review* 72, no. 1 (1992): 23–46; K. Ruggiero, 'Fingerprinting and the Argentine Plan for Universal Identification in the Late Nineteenth and Early Twentieth Centuries', in *Documenting Individual Identity: The Development of State Practices in the Modern World*, ed. Jane Caplan and John Torpey (Princeton University Press, 2001), 184–96; Julia Rodriguez, 'South Atlantic Crossings: Fingerprints, Science, and the State in Turn-of-the-Century Argentina', *The American Historical Review* 109, no. 2 (2004): 1–42.

[53] T. Herzog, *Defining Nations: Immigrants and Citizens in Early Modern Spain and Spanish America* (New Haven: Yale University Press, 2003).

[54] *Ibid.*    [55] Rodriguez, 'South Atlantic Crossings'.

descent – who in 1885 established one of the first police laboratories outside of France that used Bertillon's anthropometric methods of identification in the Argentine capital.[56] From the start, however, these efforts to use Bertillon's unspeaking *portrait parlé* to record and fix the identities of Argentines faced fierce opposition in the courts.[57]

In the 1890s Vucetich turned his attention to Galton's claims that fingerprinting was more accurate and more practical than the measurements of anthropometry. And he developed his own practical system for large-scale print classification, something that eluded Galton until Edward Henry brought one back from Bengal. Vucetich's system of classification was adopted in much of Europe and throughout Latin America, and it became the basis of a fiercely partisan effort to sustain the special achievements of fingerprinting in Argentina. By 1906 Vucetich had persuaded the Argentine police to adopt his system and a few years later the national immigration service began to include fingerprinting in the registration of arrivals. Immigrants were given a book on arrival which included instructions on the duties of citizenship, a description and a fingerprint.[58]

To this point the Argentine fingerprinting effort closely resembled the pattern used in the United States after 1937, focusing on immigrants and alleged criminals. But in 1916 Vucetich and his supporters persuaded the provincial legislature to introduce a general register of fingerprint identification, the first of its kind in the world. The courts responded quickly, declaring the new law unconstitutional and ordering the destruction of the fingerprint files that had been collected to that date.[59] Vucetich's disciple, Luis Reyna Almandos, did not give up, and decades later he attempted to develop a 'national registry of population for purposes of crime prevention' drawing on existing fingerprint repositories for military recruits and registered voters.[60] But the actual approval for a centralised population register would have to wait for decades, and it was only finally announced shortly after the 1966 military coup.[61] And even this decision had little effect on the basic patterns of civil registration in Argentina. By 2011 the national Citizen Registry contained some fifteen million fingerprint records (about the same size as the largest police database) out of

[56] Ruggiero, 'The Argentine Plan for Universal Identification', 186.
[57] Cole, *Suspect Identities*, 32–67; Ruggiero, 'The Argentine Plan for Universal Identification', 186–7.
[58] Rodriguez, 'South Atlantic Crossings', 35.
[59] Ruggiero, 'The Argentine Plan for Universal Identification', 192; Sengoopta, *Imprint of the Raj*, 192.
[60] Almandos, 'Identification in the Argentine Republic', 1098–9.
[61] Argentina, *Identification, Registration and Classification of National Human Potential*, 1968, www.mininterior.gov.ar/tramites/dni/archivos_normativas/Ley_17671.pdf.

a total population of forty million, and civil registration was still being handled by separate provincial registries organised around municipalities, and without any means of sharing data.[62] The comparison with the South African state, with its single National Population Register, no role for the municipalities, and universal fingerprint coverage, is marked.

There are important elements common to both histories. The first was the central role played by ideas of social hygiene and racial well-being in the effort to establish compulsory fingerprint registration. Progressivism in both places sought to address the problems of very rapid urban growth by applying the strong medicine of racial science.[63] And both countries have used the politics of official documentation to deny their indigenous populations the basic rights of citizenship.[64] Argentina has also served, in the distant and recent past, as an incubator for biometric systems in other Latin American countries.[65] But the real difference between the two countries lies in the Latin American emphasis on civil law, which, like the French *état civil*, places municipal civil registration at the heart of the legal order.[66] The architecture of municipal civil registration as Rosental shows for nineteenth-century France was profoundly resistant to centralisation.[67] No similar tradition existed in South Africa.

---

[62] Pedro Janices, 'Biometrical Latin America' (PowerPoint presented at the ANSI/NIST-ITL Standard 2011 Workshop, Gaithersburg, MD, 1 March 2011), www.nist.gov/itl/iad/ig/ansi_workshop-2011.cfm.

[63] Rodriguez, 'South Atlantic Crossings', 21; Zimmermann, 'Racial Ideas and Social Reform', 45.

[64] Gaston Gordillo, 'The Crucible of Citizenship: ID-Paper Fetishism in the Argentinean Chaco', *American Ethnologist* 33, no. 2 (21 April 2006): 162–76.

[65] Janices, 'Biometrical Latin America'.

[66] Phanor James Eder, Robert Joseph Kerr and Joseph Wheless, *The Argentine Civil Code (effective January 1st, 1871): Together with Constitution and Law of Civil Registry* (Boston: Boston Book, 1917).

[67] Paul-André Rosental, 'Civil Status and Identification in Nineteenth-Century France: A Matter of State Control?', in *Registration and Recognition: Documenting the Person in World History*, ed. Keith Breckenridge and Simon Szreter, Proceedings of the British Academy 182 (Oxford University Press, 2012), 137–65.

# Conclusion

The South African state was not unique in its obsessions with biometric registration but it was distinctively shaped – much more than many excellent social historians have noticed – by the goal of centralised biometric identification. And this obsession has been played out on a global stage. Like Hannah Arendt and C. Vann Woodward and many others[1] before me, in this book I have worked to explore the ways in which the institutions that developed in South Africa have served as incubators for forms of racist government that have since been globalised.

Francis Galton was a key figure in this story. The significance of his role has been obscured by the fact that his biology was eclipsed almost as soon as it became popular, that Herbert Spencer's anthropology – despite its significance a century ago – is now practically unknown, and that eugenics never secured institutional support in South Africa. Yet it is important to recognise that Galton was a figure of South African history, and that his eugenics was powerfully shaped by a racial imaginary formed by nineteenth-century South African history and its cultural politics. Like many of the new southern African histories of science, which recover the significance of vernacular scientific practices for global scholarship, this book shows that – long before the national efficiency crisis – Galton's nineteenth-century ethnography was more important than the existing international scholarship of statistics, eugenics or Darwinism has allowed. But, unlike some of the more recent studies, which have found a useable past in histories of this kind[2] this book offers, at best, a very ambivalent assessment. Whatever the current benefits

---

[1] Hannah Arendt, *The Origins of Totalitarianism* (San Diego: Harcourt Brace and Company, 1951); C. Vann Woodward, *The Strange Career of Jim Crow* (New York: Oxford University Press, 1966); John W. Cell, *The Highest Stage of White Supremacy: The Origins of Segregation in South Africa and the American South* (Cambridge University Press, 1982); Saul Dubow, *Illicit Union: Scientific Racism in Modern South Africa* (Johannesburg: Witwatersrand University Press, 1995).

[2] Saul Dubow, *A Commonwealth of Knowledge: Science, Sensibility, and White South Africa, 1820–2000* (Oxford and New York: Oxford University Press, 2006); Patrick Harries, *Butterflies & Barbarians: Swiss Missionaries & Systems of Knowledge in South-East Africa* (London: James Currey Publishers, 2007); William Beinart, Karen Brown and Daniel

of biometric registration – and they are many – Galton's writings were very destructive in their direct effects in Britain on Darwin and Spencer, and in their indirect influences in South Africa.

The biometric state was never a uniquely South African effort. Networks of empire intersected thickly on the Witwatersrand between 1895 and 1914, and without them the entire trajectory of biometric registration would have been inconceivable. It was the ambitions of imperial progressivism – unconstrained by a democratic order or constitutional traditions of local and central government – that set in place the foundations of biometric registration. What made the century-long effort possible – with the state in every generation seeking to renew the project of compulsory centralised fingerprint registration – was a particular distribution of power supported by a global racial order. So too the technologies and ideologies of compulsory biometric registration – and, perhaps even more importantly, the protests against them – echoed away from South Africa over the course of the century. Even the final collapse of Apartheid was substantially driven by global political movements that targeted the coercive forms of registration and imprisonment that distinguished South Africa. There is a sweet and perplexing irony in the fact that those same coercive systems are now being championed as the only viable remedy to the entrenched forms of poverty that are characteristic of life in the former colonies. This history should, at least, give us some reason to be cautious about the idea that biometric registration can remedy the weaknesses (and ambitions) of the post-colonial state.

The South African biometric state was also much less capable, much less interested or knowledge-driven than intellectual historians (and contemporary commentators) have often assumed. It was – like its northern neighbours – a gatekeeper state: defined by its control of the flow of resources, fiercely delimiting the transfers of benefits, constitutionally disinterested in and incapable of knowledge about the vast majority of its subjects. In South Africa this meant meticulous control of the labour districts, the policing of work-seekers as they entered the cities, careful biographical registration of white (and eventually Indian and Coloured) citizens, but comprehensive disregard for Africans in the countryside – especially for women and children. In short, the South African state was shaped by a racially defined limit on the will to know. It was this very limited state that has created the opportunities, indeed the necessity, for biometric government in the present.

Gilfoyle, 'Experts and Expertise in Colonial Africa Reconsidered: Science and the Interpenetration of Knowledge', *African Affairs* 108, no. 432 (7 January 2009): 413–33; Helen Tilley, *Africa as a Living Laboratory: Empire, Development, and the Problem of Scientific Knowledge, 1870–1950* (University of Chicago Press, 2011).

The current systems of biometric identification and cash transfer work much more efficiently than the paper-based systems they have replaced. But the mistakes and catastrophes were not confined to the old systems of paper-based fingerprinting – as the decade-long debacle of the recent HANIS project amply demonstrated.[3] Indeed, and this is the main point, to achieve the current levels of workable efficiency South Africans have had to endure a century of error. If many of the technical problems of biometric government are, mostly, now resolved in South Africa and elsewhere, it is important to remember that black and brown South Africans have paid a very high price for that result. South Africa has served as a global laboratory of biometric government but the price of the experiments was extracted from the poor over the course of the last century and they continue to pay today in the absence of institutions and infrastructure. Biometric systems for cash transfers are, in this sense, a matter of compensation, and of justice.

For the moment, the tempo of biometric government in Africa, Latin America and southern Asia continues to accelerate. Almost every country on the African continent has purchased a large-scale system of biometric identification in the last decade. Two of the most elaborate of these systems, the Nigerian and Ghanaian projects, are explicitly modelled on the South African biometric 'identification economy'.[4] And there is an appropriate symmetry in the fact that the success, or failure, of the current Unique ID project in India will determine whether the other postcolonial states follow the direction of the African states. One reason for these Southern developments was Galton's original proposal for fingerprinting as a technology of imperial government; another is that biometric government is very difficult to implement and fraught with error and inconvenience which is difficult to sustain in a vigorous democracy;

---

[3] Keith Breckenridge, 'The Elusive Panopticon: The HANIS Project and the Politics of Standards in South Africa', in *Playing the ID Card: Surveillance, Security and Identity in Global Perspective*, ed. Colin Bennett and David Lyon (London: Routledge, 2008), 39–56.

[4] In Ghana the Central Bank has purchased the world's first biometric money infrastructure from the South African firm Net1 UEPS. This new interbank switch and hundreds of thousands of smartcards encoded using their owners' fingerprints has equipped Ghana with the world's first biometric money supply. This system, which is owned by the Central Bank but designed and run by Net1 UEPS, incorporates extremely fine-grained surveillance of the monetary transactions of all of its users. Keith Breckenridge, 'The World's First Biometric Money: Ghana's E-Zwich and the Contemporary Influence of South African Biometrics', *Africa: The Journal of the International African Institute* 80, no. 4 (2010): 642–62; Keith Breckenridge, 'Capitaliser Sur Les Pauvres: Les Enjeux de L'adoption de Services Financiers Biométriques Au Nigeria', in *L'identification Biométrique: Champs, Acteurs, Enjeux Et Controverses* (Paris: Éditions de la Maison des sciences de l'homme, 2011).

yet the most important is that computerised biometric systems, notwith-standing their limits, can remedy many of the special problems of the state in Africa. In Africa, the South African bureaucratic model has been key to this process (as it was to the old forms of the gatekeeper state). But that story is nowhere near its ending.

For, despite the overblown claims that Gandhi made for satyagraha, another key finding of this history is that resistance works. Registration is very difficult at the best of times, it is practically impossible if it is deter-minedly resisted. (The painful point that Arendt made in *Eichmann in Jerusalem.*[5]) This is especially true if the resistance is waged internation-ally. The forms of global mobilisation that Gandhi used were import-ant in limiting the ambitions of biometric government in South Africa, and in restricting their use in other societies. But it is also important to keep in mind, unlike Gandhi's claims about the powers of satyagraha, once biometric registration has been offered it cannot – easily – be with-drawn. There is, outside of the liberal democracies of the North, little sign of the kinds of determined opposition that biometric registration faced in South Africa in 1907 or, again, in the early 1960s. That has much to do with the limited forms of identification that exist in the former colonies.

What distinguishes all of the biometric projects of the post-colonial world is their explicit goal to provide bureaucratic and financial ser-vices to an illiterate population. This has important theoretical impli-cations. Whatever the similarities and connections between biometrics and documentary bureaucracy, it is important to notice that biometric technologies are fundamentally – indeed ontologically – antithetical to writing. In Africa, notwithstanding the short history of literacy, there is a rich historiography and anthropology of the efforts that ordinary people have made to master the written word.[6] Biometric systems have been designed, from the early years of the twentieth century, to defeat efforts like this. My hope is that the implications of this project of transcend-ing writing might encourage scholars to reconsider the politics that is frequently attributed to literacy and the documentary state. Writing, far from being the instrument that has produced biometric government, is the only meaningful tool available to remedy its shortcomings.

---

[5] Hannah Arendt, *Eichmann in Jerusalem: A Report on the Banality of Evil* (Harmondsworth: Penguin, 2006), 135–50.

[6] See the essays in Karin Barber, *Africa's Hidden Histories: Everyday Literacy and Making the Self* (Bloomington: Indiana University Press, 2006); Derek R. Peterson, *Creative Writing: Translation, Bookkeeping, and the Work of Imagination in Colonial Kenya* (Portsmouth, NH: Heinemann, 2004).

It is also important to keep in mind that the biometric obsession was there at the beginning, and that the current administrative weakness of African states was, at least in part, a product of the models of government it produced. The limited ambitions of the gatekeeper state on the African continent cannot be entirely attributed to the precedents, ideologies and institutions that developed in South Africa, but they certainly can be traced back there in significant part. Biometric government has been partly responsible for the pitifully weak forms of local government infrastructure that are normal across the continent, and it is not likely that they will do very much to remedy them. The most we can hope for here is that the new forms of biometric citizenship, which may equip citizens with actionable entitlements, will support further demands for the other basic infrastructures like housing, clean water, sewerage systems, health-care, policing and education. Building up those systems on the African continent – but also in Latin America and in India – is likely to remain a glacial and fraught process.

Nor should we lose sight of the implications of the imperial history in the development of these new forms of citizenship. The world is already clearly divided between those subjected to centralised biometric registration systems and those who are not. This may have political implications on the borders between states in each of these zones – as anyone who has had to apply for a visa to visit Europe will attest – but a more urgent and significant problem lies in the absence of a body of law and practice that will govern biometric registration in the former colonies, and transnationally. This is especially the case where, as it often happens, the registration and data processing is being undertaken by firms with a clear commercial interest in exploiting the information.[7]

There may be some surprise, I suspect, in the realisation that an elaborate cybernetic order, very like those that Wiener and Habermas have each differently worried about, is taking form in the poorest regions of the world. This is, already, a feedback state – which Norbert Wiener worried about in the 1940s and Habermas wrote about as 'electronic steering mechanisms' – one that is powerfully shaped by what we can usefully describe as neo-Galtonian procedures of social sorting, distinguishing, in the first instance, the creditworthy from the delinquent, and where the engagement between the state and corporations (especially banks) will be driven by traces and transactions of which the subjects will be unaware. The data will be mathematically assembled, and assessed using now well-established systems of creditworthiness.

[7] Staff Writer, 'CPS in Loan Probe', *ITWeb Business*, 27 May 2013, www.itweb.co.za/index.php?option=com_content&view=article&id=64399;CPS-in-loan probe&catid=69.

Biometric registration has always had as its *raison d'être* the identification of those who cannot write. In the process it has clearly contributed to weakening the political agency of those who can write. This was Gandhi's primary worry and it persisted long after he left South Africa. Another important point – which Gandhi later disowned in *Hind Swaraj* – was that education ought to provide some leverage over the working of citizenship. Why else would we bother to be educated, he asked. Whatever may have been written about the hierarchical effects of written bureaucracy, education and literate citizenship have worked together in a virtuous partnership in many parts of the world and over centuries. The literary skills required for written registration systems allow educated citizens to insert their concerns into the workings of the bureaucracy. It was also around writing that the carefully elaborated methods of privacy protection created and preserved a zone of uncertainty around personhood. It was in that space that individuals controlled the terms and meaning of personal identity. This has been very important in the countries that have refused biometric identification. The opposite is likely to be the case in the countries that adopt the South African model of citizenship – a state-form where privacy has virtually no purchase at all, and where the citizens and officials will have few of the technical skills they require to temper or direct the trajectories of registration.

One possible solution to the political problems posed by the biometric state, a state that will leave no familiar textual traces, no paper trail at all, is to encourage the development of a proportionately large and elaborate public process of documentation, matching the state's documentary silence with a profusion of publicly accessible documents. Indeed, this is exactly what the Internet already provides. The web, as computer scientists have been saying for some time, is the first civilizational archive, and it presents exactly the sociological features of the largest repositories: it is disorderly, combining multiple, and conflicting, datasets; it is populated, in almost equal measure, by deceitful and dependable documents; and it is shaped by the plans and eccentricities of earlier generations. The great profusion of decentralised publishing that the Internet has encouraged, and the evidentiary trial that electronic mail can provide, may yet go a long way to weaken the veiled authority of the biometric state.

# Bibliography

'3 Million Must Still Get Books of Life'. *Pretoria News*, 18 April 1978.

Abendroth, Keith. 'Non-Racial ID Document Is Considered'. *The Citizen*, 15 October 1990.

Abrams, Philip. 'Notes on the Difficulty of Studying the State'. *Journal of Historical Sociology* 1, no. 1 (1988): 58–89.

Act Director-General, Department of Home Affairs. 'Not Racial ID', 1 November 1990.

Acting Secretary for Interior. Letter to Secretary for Native Affairs, Pretoria, 6 July 1911. SAB NTS 9363 4/382 Natal and Zululand: Registration of births and deaths, 1911–1960.

Acting Secretary to the Law Department. 'Finger Prints as Evidence of Identification', 13 March 1905. TAD LD 1035 1193–05 Finger prints as evidence of identification 1905.

Acting Under-Secretary for the Interior. Letter to Dower, Under Secretary for Native Affairs, 3 March 1911. SAB NTS 9363 3/382 Transkeian territories: Registration of births and deaths, 1911–1960.

Adas, Michael. *Machines as the Measure of Men: Science, Technology and Ideologies of Western Dominance*. Ithaca: Cornell University Press, 1989.

Addison, R.H. Acting Chief Native Commissioner, Natal. Letter to Secretary for Native Affairs, Pretoria. 'System of Reporting Births and Deaths, Natal Natives', 24 February 1913. SAB NTS 9363 4/382 Natal and Zululand: Registration of births and deaths, 1911–1960.

African Affairs Correspondent. 'Health and Welfare Get Priority in KwaZulu'. *The Natal Mercury*, 29 May 1979.

Agamben, Giorgio. *Homo Sacer: Sovereign Power and Bare Life*. Berkeley: University of California Press, 1998.

*State of Exception*. University of Chicago Press, 2005.

Agar, Jon. *The Government Machine: A Revolutionary History of the Computer*. Cambridge, MA: MIT Press, 2003.

'Modern Horrors: British Identity and Identity Cards'. In *Documenting Individual Identity: The Development of State Practices in the Modern World*, edited by Jane Caplan and John Torpey, 101–20. Princeton University Press, 2001.

Ake, Claude. 'The African Context of Human Rights'. *Africa Today* 34, no. 1/2 (1 March 1987): 5–12.

'Rethinking African Democracy'. *Journal of Democracy* 2, no. 1 (1991): 32–44.

Akhalwaya, Ameen. 'Storm over Govt's "Pass" Plan for All'. *Rand Daily Mail*, 15 January 1981.

'Almal Gelyk Met Vingerafdrukke'. *Die Vaderland*, 2 July 1986.

Almandos, L.R. 'Identification in the Argentine Republic'. *Journal of Criminal Law and Criminology (1931–1951)* 24, no. 6 (1934): 1098–101.

Aly, Gotz and Karl Heinz Roth. *The Nazi Census: Identification and Control in the Third Reich*. Philadelphia: Temple University Press, 2000.

Amoore, Louise. 'Biometric Borders: Governing Mobilities in the War on Terror'. *Political Geography* 25, no. 3 (March 2006): 336–51.

Anderson, Clare. *Legible Bodies: Race, Criminality, and Colonialism in South Asia*. Oxford and New York: Berg, 2004.

Anderson, Perry. *Lineages of the Absolutist State*. London: Verso, 1974.

Anderson, Ross, Ian Brown, Terri Dowty, Philip Inglesant, William Heath and Angela Sasse. *Database State*. York: Joseph Rowntree Reform Trust, 2009.

Andersson, Charles John. *Lake Ngami: Or, Explorations and Discoveries during Four Years' Wanderings in the Wilds of Southwestern Africa*. London: Dix, Edwards & Co., 1857.

Arendt, Hannah. *Eichmann in Jerusalem: A Report on the Banality of Evil*. Harmondsworth: Penguin, 2006.

   *The Origins of Totalitarianism*. San Diego: Harcourt Brace and Company, 1973.

Argentina. *Identification, Registration and Classification of National Human Potential*, 1968. www.mininterior.gov.ar/tramites/dni/archivos_normativas/Ley_17671.pdf.

Arteaga, Nelson. 'Biological and Political Identity: The Identification System in Mexico'. *Current Sociology* 59, no. 6 (1 November 2011): 754–70.

Ashurst, Mark. 'First World Smartcards and Third World Pensioners'. *Financial Times (London)*, 16 February 1996.

   'South Africa's Banking Revolution: South Africa's Remote Rural Regions Have Become an Important Testing Ground for New Banking Technology, such as Mobile ATMs and Biometrics'. *The Banker*, 1 October 1998.

Assistant Colonial Secretary. Letter to Secretary to the Law Department. 'Letter to Secretary to the Law Department', 19 June 1907. LD 1466 AG2497/07 Finger Impressions, 1907. TAD.

Aus, J.P. *Decision-Making under Pressure: The Negotiation of the Biometric Passports Regulation in the Council*. ARENA Working Paper, 11, 2006.

   'Eurodac: A Solution Looking for a Problem?' *European Integration Online Papers (EIoP)* 10, no. 6 (2006).

Bakhtin, M.M. *Dialogic Imagination: Four Essays*. Austin: Texas University Press, 1981.

Bamford, James. *Body of Secrets: How America's NSA and Britain's GCHQ Eavesdrop on the World*. London: Arrow, 2002.

   *The Puzzle Palace: A Report on America's Most Secret Agency*. London: Penguin, 1983.

Banerjee, Abhijit, Sebastian Galiani, Jim Levinsohn, Zoë McLaren and Ingrid Woolard. *Why Has Unemployment Risen in the New South Africa*. Working Paper. National Bureau of Economic Research, June 2007. www.nber.org/papers/w13167.

Barber, Karin. *Africa's Hidden Histories: Everyday Literacy and Making the Self.* Bloomington: Indiana University Press, 2006.

Bayart, Jean-Francois. *The State in Africa: The Politics of the Belly.* London: Longman, 1993.

Beinart, W. *The Rise of Conservation in South Africa: Settlers, Livestock, and the Environment 1770–1950.* New York: Oxford University Press, 2008.

Beinart, William and Colin Bundy. 'A Voice in the Big House: The Career of Headman Enoch Mamba'. In *Hidden Struggles in Rural South Africa: Politics and Popular Movements in the Transkei and Eastern Cape, 1890–1930,* 78–105. Cape Town: James Currey, 1987.

Beinart, William, Karen Brown and Daniel Gilfoyle. 'Experts and Expertise in Colonial Africa Reconsidered: Science and the Interpenetration of Knowledge'. *African Affairs* 108, no. 432 (7 January 2009): 413–33.

Beniger, James R. *The Control Revolution: Technological and Economic Origins of the Information Society.* Cambridge, MA: Harvard University Press, 1986.

Bennett, Colin J. and David Lyon. *Playing the Identity Card: Surveillance, Security and Identification in Global Perspective.* 1st edn. New York: Routledge, 2008.

Bennett, Colin J. and Charles D. Raab. *The Governance of Privacy: Policy Instruments in Global Perspective.* Aldershot: Ashgate, 2003.

Bera, Saumendranath. 'Confronting the Colonial Communication Order: Gandhiji, the Green Pamphlet and Reuter'. In *Webs of History: Information, Communication, and Technology from Early to Post-Colonial India,* edited by Amiya Kumar Baghi, Barnita Bagchi and Dipankar Sinha, 209. New Delhi: Indian History Congress, 2005.

Berry, Sara. 'Hegemony on a Shoestring: Indirect Rule and Access to Agricultural Land'. *Africa: Journal of the International African Institute* 62, no. 3 (1992): 327–55.

*No Condition Is Permanent: The Social Dynamics of Agrarian Change in Sub-Saharan Africa.* Madison: University of Wisconsin Press, 1993.

Bhana, Surendra and Goolam Vahed. *The Making of a Political Reformer: Gandhi in South Africa, 1893–1914.* New Delhi: Monahar, 2005.

Bhorat, H. and M. Oosthuizen. 'Employment Shifts and the "Jobless Growth" Debate'. In *Human Resources Development Review,* 50–68. Pretoria: HSRC Press, 2008. www.hsrcpress.ac.za/downloadpdf.php?pdffile=files%2FPDF%2F2218%2FHRD_Review.pdf&downloadfilename=Human%20Resources%20Development%20Review%202008#page=90.

'Bid to Pay Pensions Monthly'. *The Natal Mercury,* 29 March 1989.

Boje, Valerie. 'How Millions of Files Reflect SA's Reforms'. *Pretoria News,* 27 November 1990.

Bourdieu, Pierre. 'Rethinking the State: Genesis and Structure of the Bureaucratic Field'. In *State/Culture: State-Formation after the Cultural Turn,* edited by George Steinmetz, 53–75. Ithaca: Cornell University Press, 1999.

Bowcott, Owen. 'FBI Wants Instant Access to British Identity Data'. *Guardian,* 15 January 2008. www.guardian.co.uk/uk/2008/jan/15/world.ukcrime.

Bowker, Geoffrey and Susan Leigh Starr. *Sorting Things Out: Classification and Its Consequences.* Cambridge, MA: MIT Press, 1999.

Brantlinger, Patrick. 'A Postindustrial Prelude to Postcolonialism: John Ruskin, William Morris, and Gandhism', *Critical Inquiry* 22, no. 3 (Spring 1996): 466–85.

'Brazilian Bank CAIXA Deploys Lumidigm Fingerprint Sensors in ATMs', 23 July 2012. www.lumidigm.com/brazilian-bank-caixa-deploys-lumidigm-fingerprintsensors-in-atms/.

Breckenridge, Keith. 'The Biometric State: The Promise and Peril of Digital Government in the New South Africa'. *Journal of Southern African Studies* 31, no. 2 (2005): 267–82.

'Capitaliser Sur Les Pauvres: Les Enjeux de L'adoption de Services Financiers Biométriques Au Nigeria'. In *L'identification Biométrique: Champs, Acteurs, Enjeux Et Controverses*, 177–92. Paris: Éditions de la Maison des sciences de l'homme, 2011.

'The Elusive Panopticon: The HANIS Project and the Politics of Standards in South Africa'. In *Playing the ID Card: Surveillance, Security and Identity in Global Perspective*, edited by Colin Bennett and David Lyon, 39–56. London: Routledge, 2008.

'Gandhi's Progressive Disillusionment: Thumbs, Fingers, and the Rejection of Scientific Modernism in Hind Swaraj'. *Public Culture* 23, no. 2 (1 April 2011): 331–48.

'No Will to Know: The Rise and Fall of African Civil Registration in 20th Century South Africa'. In *Registration and Recognition: Documenting the Person in World History*, edited by Keith Breckenridge and Simon Szreter, 357–85. Proceedings of the British Academy 182. Oxford University Press, 2012.

'Verwoerd's Bureau of Proof: Total Information in the Making of Apartheid.' *History Workshop Journal*, 59 2005: 83–109.

'The World's First Biometric Money: Ghana's E-Zwich and the Contemporary Influence of South African Biometrics'. *Africa: The Journal of the International African Institute* 80, no. 4 (2010): 642–62.

Breckenridge, Keith and Simon Szreter, eds. *Registration and Recognition: Documenting the Person in World History*. Proceedings of the British Academy 182. Oxford University Press, 2012.

Brière, Bénédicte de la and Kathy Lindert. *Reforming Brazil's Cadastro Único to Improve the Targeting of the Bolsa Família Program*. Social Protection Discussion Paper Series. Washington, DC: Social Protection Unit, Human Development Network, the World Bank, June 2005. www.mef.gob.pe/contenidos/pol_econ/documentos/Bolsa_de_Familia_Brasil.pdf.

'Britain's Police Balk at Plug-in to FBI Database'. *Washington Times*, 16 January 2008. www.washingtontimes.com/news/2008/jan/16/britains-police32balk-at-plug-in32to-fbi-database/.

'British Association at Newcastle – Sectional Reports Continued'. *The Reader*, 19 September 1863.

*British East Africa. Offer of Appointment as Finger Print Expert to LieutenantWcc Burgess Transmits Information Regarding Conditions of Service*. Vol. 774. GG, 1919.

British Indian Association. 'Petition to Secretary of State for Colonies', 9 September *British East Africa* 1908. Collected Works, Volume 9, 119. Collected Works.

Brits, E. 'Kwessie van Ras en Vingerafdrukke Pla Nog', *Die Burger*, 15 June 1990.

Brookes, Edgar. *The History of Native Policy in South Africa from 1830 to the Present Day*. Cape Town: Nasionale Pers, 1924.

Brown, Judith Margaret. *Gandhi: Prisoner of Hope*. New Haven: Yale University Press, 1991.

Browne, D.G. *The Rise of Scotland Yard*. London: George G. Harrap, 1956.

Bucknill, John A. Letter to Attorney General. 'Report of a Departmental Enquiry Held by Mr John A Bucknill into Certain Complaints Made by the Commissioner of Police against the Acting Chief Detective Inspector on His Establishment and into Other Matters in Connection Therewith', 7 May 1906. TAD C30 11 PSC 454/06 Mr Brucknill's report on certain police matters 1906.

Bull, Malcolm. 'States Don't Really Mind Their Citizens Dying (Provided They Don't All Do It at Once): They Just Don't Like Anyone Else to Kill Them'. *London Review of Books* 26, no. 24 (16 December 2004): 3–6.

Burchell, Peter, Colin Gordon and Peter Miller, eds. *The Foucault Effect: Studies in Governmentality with Two Lectures by and an Interview with Michel Foucault*. London: Harvester Wheatsheaf, 1991.

Bureau of Census and Statistics, South Africa. *Report of the Director, 1951*. Pretoria, 1951.

*Report of the Director, 1952*. Pretoria, 1952.

*Report of the Director, 1957*. Pretoria, 1957.

Burley, Henry. Letter to Registrar of Asiatics, Department of the Interior. 'Report on the Workings of the Fingerprint System', 22 April 1912. CAD JUS 0862, 1/138. URL.

Calder, James D. *The Origins and Development of Federal Crime Control Policy: Herbert Hoover's Initiatives*. Westport: Praeger, 1993.

Caplan, Jane. 'Registering the Volksgemeinschaft. Civil Status in Nazi Germany 1933–1939'. In *A Nazi 'Volksgemeinschaft'? German Society in the Third Reich*, 116–28. German Historical Institute London and Oxford University Press, 2012.

'"This or That Particular Person": Protocols of Identification in Nineteenth-Century Europe'. In *Documenting Individual Identity: The Development of State Practices in the Modern World*, edited by Jane Caplan and John Torpey, 49–66. Princeton University Press, 2001.

Caplan, Jane and John C. Torpey. *Documenting Individual Identity: The Development of State Practices in the Modern World*. Princeton University Press, 2001.

Carlyle, Thomas. 'Occasional Discourse on the Negro Question'. *Fraser's Magazine for Town and Country*, February 1849. http://homepage.newschool.edu/het//texts/carlyle/carlodnq.htm.

Carton, Benedict. *Blood from Your Children: The Colonial Origins of Generational Conflict in South Africa*. Pietermaritzburg: University of Natal Press, 2000.

Cashmore, Simon. 'Datakor: Digesting Unisys'. *Financial Mail*, 17 August 1990.

Cash Paymaster Services. 'Questionnaire: Computerworld Smithsonian Awards'. Computerworld Smithsonian Awards, 1995. Archives Center at American History, Smithsonian Institution.

Castells, Manuel. *End of Millenium*. 3 vols. Vol. 3. Oxford: Blackwell, 1998.

*The Power of Identity*. 3 vols. Vol. 2. 2nd edn. Oxford: Wiley Blackwell, 1997.

*The Rise of the Network Society*. Oxford: Blackwell, 1996.

Cate, Fred H. *Privacy in the Information Age.* Washington, DC: Brookings Institution Press, 1997.

Cell, John W. *The Highest Stage of White Supremacy: The Origins of Segregation in South Africa and the American South.* Cambridge University Press, 1982.

Chamberlain, Secretary of State, London. Letter to Milner, High Commissioner, Cape Town, 18 July 1900. TAD PSY 1 HC102/00 High Commissioner: Johannesburg Police Henry Appointed Temporary 1900–1900.

Chamney, Montfort. 'Mahatma Gandhi in the Transvaal', 1935. Mss Eur C859 1902–1914. British Library, Asia, Pacific and Africa Collections. www.nationalarchives.gov.uk/a2a/records.aspx?cat=059-msseur_2&cid=-1#-1.

Chamney, Protector and Registrar of Asiatics. Letter to Lionel Curtis, Assistant Colonial Secretary (Division II). 'Report on the Position of Asiatics in the Transvaal (irrespective of Chinese Indentured Labour) in Relation Especially to Their Admission and Registration', 17 April 1906. TAD, LTG 97, 97/03/01 Asiatics Permits, 1902–7. TAD.

Chamney, Registrar of Asiatics, Department of the Interior. Letter to Acting Secretary for Justice. 'Confidential', 23 April 1912. CAD JUS 0862, 1/138, 1910.

Chanock, Martin. *The Making of South African Legal Culture, 1902–1936: Fear, Favour and Prejudice.* Cambridge University Press, 2001.

Chatterjee, P. *Nationalist Thought and the Colonial World: A Derivative Discourse.* London: Zed Books, 1986.

Chertoff, Michael. 'Remarks by Secretary Michael Chertoff to the Johns Hopkins University Paul H. Nitze School of Advanced International Studies'. *Department of Homeland Security,* 3 May 2007. www.dhs.gov/xnews/speeches/sp_1178288606838.shtm.

Chief Native Commissioner, Pietermaritzburg. Letter to Secretary for Native Affairs, Pretoria. 'Registration of Native Births and Deaths', 22 December 1922. SAB NTS 9363 4/382 Natal and Zululand: Registration of births and deaths, 1911–1960.

Letter to Secretary for Native Affairs, Pretoria, September 17, 1932. SAB NTS 9363 4/382 Natal and Zululand: Registration of births and deaths, 1911–1960.

*Ciskei. Bevolkingsregistrasie.* Vol. 11/98. BAO, 1981.

Clanchy, M.T. *From Memory to Written Record, England 1066–1307.* London: Edward Arnold, 1979.

Clarke, Jack. Interview by Keith Breckenridge and Paul Edwards, 20 April 2004.

Clarke, W.J., Chief Commissioner, Natal Police. *Evidence to Natal Native Affairs Commission, 1906–7.* Pietermaritzburg: Government Printers, 1907.

Letter to Secretary, South African Police, 12 February 1913. SAB NTS 9363 4/382 Natal and Zululand: Registration of births and deaths, 1911–1960.

'Clearing up the Book of Life Confusion'. *Rand Daily Mail,* 14 December 1977.

Clinton, Hillary. 'Smart Power Approach to Counterterrorism'. John Jay School of Criminal Justice, 9 September 2011. www.state.gov/secretary/rm/2011/09/172034.htm.

Cohn, Bernard S. *Colonialism and Its Forms of Knowledge: The British in India.* Delhi: Oxford University Press, 1997.

Cole, Ernest. *House of Bondage By Ernest Cole (A South African Black Man Exposes in His Own Pictures and Words the Bitter Life of His Homeland Today)*, edited by Thomas Flaherty. New York: Random House, 1967.

Cole, Simon A. 'Digits – Automated Fingerprinting and New Biometric Technologies'. Unpublished paper, n.d.

'Is Fingerprint Identification Valid? Rhetorics of Reliability in Fingerprint Proponents' Discourse'. *Law & Policy* 28, no. 1 (January 2006): 109–35.

*Suspect Identities: A History of Fingerprinting and Criminal Identification.* Cambridge, MA: Harvard University Press, 2001.

'Witnessing Identification: Latent Fingerprinting Evidence and Expert Knowledge'. *Social Studies of Science* 28, no. 5/6, Special Issue on Contested Identities: Science, Law and Forensic Practice (1998): 687–712.

Colonial Secretary. 'NAB CSO 1893, 1910/4799 Secretary for Interior Pretoria Asks for Memorandum Shewing System Prevailing in Natal for Registration of Births and Deaths, Particularly Natives', 1910. Pietermaritzburg.

'NAB CSO 1902, 1911/1482 Secretary for the Interior Pretoria: Asks to Be Furnished with a Memorandum Under Which the Systems of Registration of Births, Marriages and Deaths Is Administered', 1911.

Comaroff, Jean and John L. Comaroff. *Of Revelation and Revolution: Christianity, Colonialism, and Consciousness in South Africa.* Vol. 2. Chicago University Press, 1997.

*Of Revelation and Revolution: The Dialectics of Modernity on a South African Frontier.* Vol. 1. University of Chicago Press, 1991.

'Comment'. *Sowetan*, 4 April 1986.

Cooper, Frederick. *Africa Since 1940: The Past of the Present.* Cambridge University Press, 2002.

*Decolonization and African Society: The Labor Question in French and British Africa.* Cambridge University Press, 1996.

'Voting, Welfare, and Registration: The Strange Fate of the Etat-Civil in French Africa, 1945–1960'. In *Registration and Recognition: Documenting the Person in World History*, edited by Keith Breckenridge and Simon Szreter, 385–414. Proceedings of the British Academy 182. Oxford University Press, 2012.

Coovadia, Hoosen, Rachel Jewkes, Peter Barron, David Sanders and Diane McIntyre. 'The Health and Health System of South Africa: Historical Roots of Current Public Health Challenges'. *The Lancet* 374, no. 9692 (5 September 2009): 817–34.

Corrigan, Philip Richard D. and Derek Sayer. *The Great Arch: English State Formation as Cultural Revolution.* Oxford: Blackwell, 1985.

Cotter, Holland. '"Rise and Fall of Apartheid" at Center of Photography'. *New York Times*, 20 September 2012. www.nytimes.com/2012/09/21/arts/design/rise-and-fall-of-apartheid-at-center-of-photography.html.

Cowan, Ruth Schwartz. 'Francis Galton's Statistical Ideas: The Influence of Eugenics'. *Isis* 63, no. 4 (1972): 509–28.

Crush, Jonathan. 'Power and Surveillance on the South African Gold Mines'. *Journal of Southern African Studies* 18, no. 4 (1992).

Crush, J.S. and Charles H. Ambler. *Liquor and Labor in Southern Africa*. Athens: Ohio University Press, 1992.

Cluver, E.H., Asst Health Officer. 'Malaria Conference Held at Durban, Town Hall, on 1st July 1932'. 4 July 1932. SAB NTS 9363 4/382 Natal and Zululand: Registration of births and deaths, 1911–1960.

Curtin, Philip D. *The Image of Africa: British Ideas and Action, 1780–1850*. 2 vols. Madison: University of Wisconsin Press, 1964.

Curtis, Lionel. Assistant Colonial Secretary (Division II). Letter to Patrick Duncan, Colonial Secretary. 'Position of Asiatics in the Transvaal', 1 May 1906. LTG 97, 97/03/01 Asiatics Permits, 1902–7. TAB.

'The Place of Subject People in the Empire', 9 May 1907. A146, Fortnightly Club, Johannesburg, South African Historical Archive.

*With Milner in South Africa*, n.d.

Daily News Reporter. 'Black Sash Slams KwaZulu Government's Pension Scheme'. *Daily News*, 30 September 1989.

'KwaZulu Accepts Pension Proposals'. *Daily News*, 7 May 1986.

Darwin, Charles. *The Descent of Man, and Selection in Relation to Sex*. London: D. Appleton, 1872.

Davin, Anna. 'Imperialism and Motherhood'. *History Workshop Journal* 5 (1978): 9–65.

De Janvry, A., F. Finan, E. Sadoulet, D. Nelson, K. Lindert, B. de la Brière and P. Lanjouw. 'Brazil's Bolsa Escola Program: The Role of Local Governance in Decentralized Implementation'. Washington, DC: World Bank, Social Protection (SP), SP Discussion Paper no. 0542 (2005). http://siteresources. worldbank.org/SOCIALPROTECTION/Resources/SP-Discussion-papers/ Safety-Nets-DP/0542.pdf.

Delius, Peter. *The Land Belongs to Us: The Pedi Polity, the Boers and the British in the Nineteenth Century Transvaal*. Johannesburg: Ravan Press, 1983.

*A Lion Amongst the Cattle: Reconstruction and Resistance in the Northern Transvaal*. Oxford: James Currey, 1996.

Denoon, Donald. *A Grand Illusion: The Failure of Imperial Policy in the Transvaal Colony during the Period of Reconstruction, 1900–1905*. London: Longman, 1973.

Department of Interior, South Africa. *Annual Report for the Calendar Year 1972*. Pretoria, 1972.

*Annual Report for the Calendar Year 1973*. Pretoria, 31 December 1973.

*Annual Report for the Calendar Year 1977*. Pretoria, 1977.

Desmond, Adrian J. and James Richard Moore. *Darwin's Sacred Cause: How a Hatred of Slavery Shaped Darwin's Views on Human Evolution*. Boston: Houghton Mifflin Harcourt, 2009.

Dhupelia-Mesthrie, Uma. *Gandhi's Prisoner: The Life of Gandhi's Son Manilala*. Cape Town: Kwela Books, 2007.

Dilnot, G. *The Story of Scotland Yard*. Boston: Houghton Mifflin Company, 1927.

Dinwiddy, J.R. 'The Early Nineteenth-Century Campaign against Flogging in the Army'. *The English Historical Review* 97, no. 383 (April 1982): 308–31.

Dower, Edward, [Acting?] Secretary for Native Affairs. Letter to Burton, Minister for Native Affairs. 'Registration of Births and Deaths of Natives in Zululand',

18 July 1911. SAB NTS 9363 4/382 Natal and Zululand: Registration of births and deaths, 1911–1960.

Secretary for Native Affairs, Cape Colony. Letter to Prime Minister, Cape Colony. 'Finger Print System – Habitual Criminals: Proposed Extension to the Transkeian Territories', 17 September 1908. SAB NA 168, 1/1908/F372 Introduction of Finger Print System into the Transkeian Territories, 1913.

Dubow, Saul. *A Commonwealth of Knowledge: Science, Sensibility, and White South Africa, 1820–2000*. Oxford and New York: Oxford University Press, 2006.

'Holding "a Just Balance between White and Black": The Native Affairs Department in South Africa c.1920–33'. *Journal of Southern African Studies* 12, no. 2 (April 1986): 217–39.

*Illicit Union: Scientific Racism in Modern South Africa*. Johannesburg: Witwatersrand University Press, 1995.

*Racial Segregation and the Origins of Apartheid in South Africa, 1919–36*. Oxford: Macmillan in association with St Antony's College, 1989.

Dugger, Celia W. 'Ernest Cole's Photographs Exhibited in South Africa'. *New York Times*, 17 November 2010. www.nytimes.com/2010/11/18/arts/design/18cole.html.

Duminy, Andrew. 'Towards Union, 1900–10'. In *Natal and Zululand from Earliest Times to 1910: A New History*, edited by Andrew Duminy and Bill Guest, 402–23. Pietermaritzburg: University of Natal Press, 1989.

Dussel, Enrique. 'Eurocentrism and Modernity (Introduction to the Frankfurt Lectures)'. *Boundary 2* 20, no. 3 (1 October 1993): 65–76.

Dvornychenko, V.N. and Michael D. Garris. *Summary of NIST Latent Fingerprint Testing Workshop*, November 2006.

Eder, Phanor James, Robert Joseph Kerr and Joseph Wheless. *The Argentine Civil Code (effective January 1st, 1871): Together with Constitution and Law of Civil Registry*. Boston: Boston Book, 1917.

Editor. 'Bewysboekstelsel'. *Die Transvaler*, 4 November 1977.

'Black Pensions'. *The Daily News*, 4 September 1982.

Editorial. 'Do We Need It?' *The Natal Mercury*, 16 January 1981.

'It's a Blot'. *The Citizen*, 17 January 1981.

'Pensions: Biometric Answers'. *Financial Mail*, 5 June 1992.

Edwards, Paul N. *The Closed World: Computers and the Politics of Discourse in Cold War America*. Cambridge, MA: MIT Press, 1996.

*A Vast Machine: Computer Models, Climate Data, and the Politics of Global Warming*. Cambridge, MA: MIT Press, 2010.

Edwards, Paul N. and Gabrielle Hecht. 'History and the Technopolitics of Identity: The Case of Apartheid South Africa'. *Journal of Southern African Studies* 36, no. 3 (September 2010): 619–39.

Eldredge, Elizabeth and Fred Morton, eds. *Slavery in South Africa: Captive Labor on the Dutch Frontier*. Pietermaritzburg: University of Natal Press, 1994.

Emmett, Tony and Gerard Hagg. 'Politics of Water Management: The Case of the Orange River Development Project'. In *Empowerment through Economic Transformation*, edited by Meshack Khosa, 299–328. Pretoria: HSRC Press, 2001.

Etherington, Norman. *The Great Treks: The Transformation of Southern Africa, 1815–1854*. London: Longman, 2001.

'The "Shepstone System" in the Colony of Natal and beyond the Borders'. In *Natal and Zululand from Earliest Times to 1910: A New History*, 170–92. Pietermaritzburg: University of Natal Press and Shuter & Shooter, 1989.

Evans, Ivan. *Bureaucracy and Race: Native Administration in South Africa*. Berkeley: University of California Press, 1997.

Fabian, Johannes. 'Hindsight. Thoughts on Anthropology Upon Reading Francis Galton's Narrative of an Explorer in Tropical South Africa (1853)'. *Critique of Anthropology* 7, no. 2 (1987): 37–49.

Fahmy, Khaled. *All the Pasha's Men: Mehmed Ali, His Army, and the Making of Modern Egypt*. Cairo: American University in Cairo Press, 1997.

Falk, Myron. 'Fingerprints: Black Marks against the Migrant'. *Social Forces* 19, no. 1 (1 October 1940): 52–6.

Falwasser, H.G., Acting Director of Native Labour. *Report of the Director of Native Labour*. Transvaal Administration, 1 November 1926. TAD GNLB 380 11/11 Native Affairs Department General, 1921–1935.

Fancher, Raymond E. 'Francis Galton's African Ethnography and Its Role in the Development of His Psychology'. *The British Journal for the History of Science* 16, no. 1 (March 1983): 67–79.

Fanon, Frantz. *The Wretched of the Earth: A Negro Psychoanalyst's Study of the Problems of Racism and Colonialism*. New York: Grove Press, 1966.

Ferguson, James. 'What Comes After the Social? Historicizing the Future of Social Assistance and Identity Registration in Africa'. In *Registration and Recognition: Documenting the Person in World History*, edited by Keith Breckenridge and Simon Szreter, 495–516. Proceedings of the British Academy 182. Oxford University Press, 2012.

Fields, Barbara J. and Karen Fields. *Racecraft: The Soul of Inequality in American Life*. London: Verso, 2012.

Finance Editor. 'KwaZulu System Sign of the Future'. *Natal Mercury*, 18 July 1990.

*Fingerprint Expert Loan of Services to Government of Sierra Leone*. Vol. 347. SAP, 1945.

'Fingerprinting SA'. *The Cape Times*, 8 April 1986.

Fischer, Louis. *The Life of Mahatma Gandhi*. New York: HarperCollins Publishers Ltd, 1997.

The Fletcher School, Tufts University and Bankable Frontier Associates. 'Is Grandma Ready for This? Mexico Kills Cash-Based Pensions and Welfare by 2012'. The Fletcher School, Tufts University, 2011.

Foucault, Michel. *Discipline and Punish: The Birth of the Prison*. London: Penguin, 1977.

'Governmentality'. In *The Foucault Effect: Studies in Governmentality with Two Lectures by and an Interview with Michel Foucault*, edited by Peter Burchell, Colin Gordon and Peter Miller, 87–104. London: Harvester Wheatsheaf, 1991.

Frankel, Philip H. *Pretoria's Praetorians: Civil-Military Relations In South Africa*. Cambridge University Press, 1984.

Freeden, Michael. 'Eugenics and Progressive Thought: A Study in Ideological Affinity'. *Historical Journal* 22, no. 3 (1979): 645–71.

F R for Secretary for Native Affairs. Letter to Director of Census and Statistics, Pretoria. 'National Register of Population', 12 October 1949. SAB NTS 9360 2/382 Part 2 Registration of Births and Deaths (natives) in the Union. (general File). 1947 to 1952.

Galton, Francis. *The Art of Travel; or Shifts and Contrivances Available in Wild Countries*. 1st edn. London: John Murray, 1855.

*Finger Prints*. London and New York: Macmillan and Co., 1892.

'Gregariousness in Cattle and in Men'. *Macmillan's Magazine*, 1871.

*Hereditary Genius: An Inquiry into Its Laws and Consequences*. London: Macmillan, 1869.

'Hereditary Talent and Character'. *Macmillan's Magazine*, 1865.

'Identification by Finger Tips'. *Nineteenth Century* 30 (August 1891): 303–11.

'Identification Offices in India and Egypt'. *Nineteenth Century* 48 (July 1900): 118–26.

*Inquiries Into Human Faculty and Its Development*. London: Macmillan, 1883.

*Memories of My Life*. London: Methuen & co., 1908.

*The Narrative of an Explorer in Tropical South Africa*. London: J. Murray, 1853.

Gandhi, M.K. 'Baroda: A Model Indian State'. *Indian Opinion*, 3 June 1905. *Collected Works*, vol. 4, 302.

'The British Indian Association and Lord Milner'. *Indian Opinion*, 11 June 1903. *Collected Works*, vol. 3, 64–73.

'A Contrast'. *Indian Opinion*, 10 March 1906. *Collected Works*, vol. 5, 113.

'A Dialogue on the Compromise'. *Indian Opinion*, 8 February 1908. *Collected Works*, vol. 8, 136–47.

'Deputation to Colonial Secretary'. *Indian Opinion*, 17 March 1906.

'Fair and Just Treatment'. *Indian Opinion*, 11 August 1906. *Collected Works*, vol. 5, 300.

*Hind Swaraj and Other Writings*, edited by A.J. Parel. Cambridge University Press, 1997.

'Indian Promissory Notes'. *Indian Opinion*, 2 July 1904. *Collected Works*, vol. 4, 28.

Interview to Nirmal Kumar Bose. Interview by Nirmal Kumar Bose, 9 November 1934. *Collected Works*, vol. 65, 316.

'Johannesburg Letter'. *Indian Opinion*, 28 December 1907.

'Johannesburg Letter'. *Indian Opinion*, 29 February 1908. *Collected Works*, vol. 8, 169.

Letter to Attorney General, Natal. 'Letter to Attorney General', 30 June 1904. MJPW 128 in MJ1287 'MK Gandhi Re bill to regulate the signing of negotiable instruments by Indians. Suggests the finger prints also be taken'. 30 June 1904.

'Letter to Colonial Secretary'. *Indian Opinion*, 7 October 1907.

Letter to Manmohandas Gandhi. 'Letter to Manmohandas P. Gandhi', 3 March 1930. *Collected Works*, vol. 48, 371.

'Letter to "The Star"', 17 September 1908.

'Mass Meeting of Transvaal Indians: Full Account'. *Indian Opinion*, 6 April 1907.

'Oppression in the Cape'. *Indian Opinion*, 29 December 1906. *Collected Works*, vol. 6, 201.

'Petition to Natal Legislative Assembly', 28 June 1894. *Collected Works*, vol. 1.

'Sarvodaya'. *Indian Opinion*, 16 May 1908. *Collected Works*, vol. 9.

'Sarvodaya [-IX]'. *Indian Opinion*, 18 July 1908. *Collected Works*, vol. 8, 455.

*Satyagraha in South Africa*. Vol. 3. *The Selected Works of Mahatma Gandhi*. Ahmedabad: Navajivan Publishing House, 1928.

'Statement by the Delegates on Behalf of the British Indians in the Transvaal Regarding the "Petition" from Dr. William Godfrey and Another and Other Matters', 20 November 1906. *Collected Works*, vol. 6, 126.

'Transvaal Asiatic Ordinance'. *Indian Opinion*, 30 March 1907.

'Trial of PK Naidoo and Others'. *Indian Opinion*, 28 December 1907. *Collected Works*, vol. 8, 41.

Gani, Abdul, Chairman, British Indian Association. Letter to Patrick Duncan. 'Letter to Colonial Secretary', 25 August 1906.

Garcia, M. and C. Moore. *The Cash Dividend: The Rise of Cash Transfer Programs in Sub-Saharan Africa*. Washington, DC: The World Bank, 2012. www-wds. worldbank.org/external/default/WDSContentServer/WDSP/IB/2012/03/01 /000386194_20120301002842/Rendered/PDF/672080PUB0EPI0020Box 367844B09953137.pdf.

Garland, D. *The Culture of Control: Crime and Social Order in Contemporary Society*. New York: Oxford University Press, 2001.

Gates, Kelly. 'The United States Real ID Act and the Securitization of Identity'. In *Playing the Identity Card: Surveillance, Security and Identification in Global Perspective*, edited by David Lyon and Colin Bennett, 218–32. London and New York: Routledge, 2008.

Geertz, Clifford. *The Interpretation of Cultures: Selected Essays*. New York: Basic Books, 1973.

Gelb, A. and C. Decker. 'Cash at Your Fingertips: Biometric Technology for Transfers in Developing Countries'. *Review of Policy Research* 29, no. 1 (2012): 91–117.

Gelb, Alan and Julia Clark. 'Identification for Development: The Biometrics Revolution'. Centre for Global Development, 28 January 2013. www.cgdev. org/content/publications/detail/1426862/.

Gevers, Liezl. 'Black Sash and Red Tape: The Plight of the African Aged in KwaZulu and Natal, 1979–1990'. Honours in History, University of KwaZulu-Natal, 2012.

Gewald, Jan-Bart. *Herero Heroes: A Socio-Political History of the Herero of Namibia, 1890–1923*. Athens, OH: Ohio State University Press, 1999.

Giddens, A. *The Nation-State and Violence*. Vol. 2. *A Contemporary Critique of Historical Materialism*. Berkeley: University of California Press, 1985.

Giliomee, Hermann. *The Afrikaners: Biography of a People*. Cape Town: Tafelberg, 2003.

Gillham, N.W. *A Life of Sir Francis Galton: From African Exploration to the Birth of Eugenics*. New York: Oxford University Press, 2001.

Ginzburg, Carlo. 'Clues: Roots of an Evidential Paradigm'. In *Clues, Myths and the Historical Method*, 96–125. Baltimore: Johns Hopkins University Press, 1989.

Glaser, D. 'Behind the Indaba: The Making of the KwaNatal Option'. *Transformation* no. 2 (1986). http://transformation.ukzn.ac.za/index.php/transformation/article/viewFile/394/213.

Goldberg, David Theo. *The Threat of Race: Reflections on Racial Neoliberalism.* Malden: John Wiley & Sons, 2009.

Goodman, Sy. 'Computing in South Africa; an End to Apartness?' *Communications of the ACM* 37, no. 2 (February 1994): 21–5.

'Good Move If'. *Daily News*, 9 November 1984.

Goody, Jack. *The Logic of Writing and the Organization of Society.* Oxford University Press, 1986.

Gordillo, Gaston. 'The Crucible of Citizenship: ID-Paper Fetishism in the Argentinean Chaco'. *American Ethnologist* 33, no. 2 (April 21, 2006): 162–176.

Gordon, Robert J. 'The Venal Hottentot Venus and the Great Chain of Being'. *African Studies* 51, no. 2 (1992): 185–201.

Gorski, Philip S. 'The Protestant Ethic Revisited: Disciplinary Revolution and State Formation in Holland and Prussia'. *The American Journal of Sociology* 99, no. 2 (1993): 265–316.

Gould, Stephen J. 'The Geometer of Race'. *Discover* (November 1994): 65–9.

Gray, G.T. and R.Q. Smith. 'Against the Current: The Sperry-Burroughs Merger and the Unisys Struggle to Survive 1980–2001'. *Annals of the History of Computing, IEEE* 29, no. 2 (June 2007): 3–17.

Groebner, Valentin. *Who Are You? Identification, Deception, and Surveillance in Early Modern Europe.* New York: Zone Books, 2007.

Grubb, Nick. 'Below-the-Belt Scam Uncovered'. *Saturday Star*, 16 November 1996.

Guha, R. 'The Prose of Counter-Insurgency'. In *Selected Subaltern Studies*, 45–84. New York: Oxford University Press, 1988.

Guild, June Purcell. 'Transients in a New Guise'. *Social Forces* 17, no. 3 (1939): 366–72.

Gupta, Akhil. 'Blurred Boundaries: The Discourse of Corruption, the Culture of Politics, and the Imagined State'. *American Ethnologist* 22, no. 2 (1 May 1995): 375–402.

Gurney, Christabel. '"A Great Cause": The Origins of the Anti-Apartheid Movement, June 1959–March 1960'. *Journal of Southern African Studies* 26, no. 1 (1 March 2000): 123–44.

Guy, Jeff. *The Heretic: A Study of the Life of John William Colenso.* Johannesburg: Ravan Press, 1983.

   *The Maphumulo Uprising: War, Law and Ritual in the Zulu Rebellion.* Pietermaritzburg: University of KwaZulu-Natal Press, 2005.

Habermas, J. *The Theory of Communicative Action. Vol. 2: Lifeworld and System: A Critique of Functionalist Reason.* Boston: Beacon Press, 1992.

Hacking, Ian. *The Taming of Chance.* Cambridge University Press, 1990.

Hall, Catherine. *Civilising Subjects: Metropole and Colony in the English Imagination 1830–1867.* Cambridge: Polity Press, 2002.

   'The Economy of Intellectual Prestige: Thomas Carlyle, John Stuart Mill, and the Case of Governor Eyre'. *Cultural Critique* no. 12 (Spring 1989): 167–96.

Hamilton, Carolyn, ed. *The Mfecane Aftermath: Reconstructive Debates in Southern African History*. Johannesburg: Witwatersrand University Press, 1995.

Hancock, W.K. *Smuts: 1 the Sanguine Years, 1870–1919*. Cambridge University Press, 1962.

'HANIS Basic System Commissioning'. 18 February 2002. www.info.gov.za/speeches/2002/0202191046a1001.htm.

Hanlon, J. 'It Is Possible to Just Give Money to the Poor'. In *Catalysing Development? A Debate on Aid*, 181. Oxford: Blackwell, 2004.

'Just Give Money to the Poor.' Conference Item, 22 April 2009. http://oro.open.ac.uk/20887/.

Hanlon, J., D. Hulme and A. Barrientos. *Just Give Money to the Poor: The Development Revolution from the Global South*. Sterling: Kumarian Press, 2010.

Harries, Patrick. *Butterflies & Barbarians: Swiss Missionaries & Systems of Knowledge in South-East Africa*. London: James Currey Publishers, 2007.

*Work, Culture and Identity: Migrant Labourers in Mozambique and South Africa, c 1860–1910*. Johannesburg: Witwatersrand University Press, 1994.

Harris, Jose. *William Beveridge: A Biography*. Rev. edn Oxford: Clarendon Press, 1997.

Hassim, Shireen. '"A Conspiracy of Women": The Women's Movement in South Africa's Transition to Democracy'. *Social Research* 69, no. 3 (Fall 2002): 693–732.

'Social Justice, Care and Developmental Welfare in South Africa: A Capabilities Perspective'. *Social Dynamics* 34, no. 2 (2008): 104–18.

Headrick, D.R. *The Tools of Empire: Technology and European Imperialism in the Nineteenth Century*. Oxford University Press, 1981.

Hegel, Georg Wilhelm Friedrich. *Elements of the Philosophy of Right*, edited by A.W. Wood. Translated by H.B. Nisbet. Cambridge University Press, 2008.

Henry, Edward. *Classification and Uses of Finger Prints*. London: G. Routledge and Sons, 1900.

Letter to Sir Alfred Milner. 'Draft Act for the Regulation of the Town Police', 11 November 1900. TAD LD 15 LAM 1150/01 Commissioner of Police Reports Transvaal Police 1901.

Letter to Sir Alfred Milner. 'Memorandum on a Police Organisation for Johannesburg and the Witwatersrand District', 18 September 1900. TAD LD 15 LAM 1150/01 Commissioner of Police Reports Transvaal Police 1901.

Letter to Fiddes, W. 'Police Organisation for the Towns of the Transvaal', 2 November 1900. TAD LD 15 LAM 1150/01 Commissioner of Police Reports Transvaal Police 1901.

Letter to Fiddes, W, 13 February 1901. TAD LD 15 LAM 1150/01 Commissioner of Police Reports Transvaal Police 1901.

Letter to General Maxwell. 'Proposal for the Policing of Pretoria', 26 November 1900. TAD LD 15 LAM 1150/01 Commissioner of Police Reports Transvaal Police 1901.

Letter to Milner, High Commissioner, Cape Town, 11 October 1900. TAD LD 15 LAM 1150/01 Commissioner of Police Reports Transvaal Police 1901.

*Police Regulations, & c: Published by Authority*. Pretoria: Government Printing Office, 1901.

Herbst, Jeff. *States and Power in Africa: Comparative Lessons in Authority and Control*. Princeton University Press, 2000.

Herschel, William James. *The Origin of Finger-Printing*. London: Oxford University Press, 1916. www.archive.org/details/originoffingerpr00hersrich.

Herzog, T. *Defining Nations: Immigrants and Citizens in Early Modern Spain and Spanish America*. New Haven: Yale University Press, 2003.

Higgs, Edward. 'Fingerprints and Citizenship: The British State and the Identification of Pensioners in the Interwar Period'. *History Workshop Journal* 69 (2010): 52–67.

    *Identifying the English: A History of Personal Identification, 1500 to the Present*. London and New York: Continuum, 2011.

    *The Information State in England: The Central Collection of Information on Citizens since 1500*. New York: Palgrave Macmillan, 2004.

Hindson, Douglas. *Pass Controls and the Urban African Proletariat*. Johannesburg: Ravan Press, 1987.

    'Pass Offices and Labour Bureaux: An Analysis of the Development of a National System of Pass Controls in South Africa'. PhD, University of Sussex, 1981.

Hobbes, Thomas. *Leviathan*. Harmondsworth: Penguin, 1985.

Hofstadter, Richard. *The Age of Reform: From Bryan to FDR*. New York: Alfred A Knopf, 1956.

Hogan, Michael J. *A Cross of Iron: Harry S Truman and the Origins of the National Security State*. Cambridge University Press, 1998.

Honnold, William L. 'The Negro in America', n.d. Fortnightly Club, Johannesburg. South African Historical Archive, University of the Witwatersrand.

Hoover, J. Edgar. 'Criminal Identification'. *The American Journal of Police Science* 2, no. 1 (1931): 09/08/19.

Hughes, Thomas. 'Technological Momentum'. In *Does Technology Drive History? The Dilemma of Technological Determinism*, edited by Michael L. Smith and Leo Marx, 101–13. Cambridge, MA: MIT Press, 1994.

Hull, Matthew S. 'Documents and Bureaucracy'. *Annual Review of Anthropology* 41 (2012): 251–67.

    'Ruled by Records: The Expropriation of Land and the Misappropriation of Lists in Islamabad'. *American Ethnologist* 35, no. 4 (2008): 501–18.

Identification Branch, Foreign Labour Department. 'Report on the Work of the Identification Branch of the Foreign Labour Department', 1 September 1905. TAD FLD 171, 35.

IIVRS. *The Impact of Computerization on Population Registration in Sweden*. Technical Papers. Bethesda: International Institute for Vital Registration and Statistics, December 1996.

Janices, Pedro. 'Biometrical Latin America'. PowerPoint presented at the ANSI/NIST-ITL Standard 2011 WORKSHOP, Gaithersburg, MD, 1 March 2011. www.nist.gov/itl/iad/ig/ansi_workshop-2011.cfm.

Jeeves, Alan H. 'The Control of Migratory Labour on the South African Gold Mines in the Era of Kruger and Milner'. *Journal of Southern African Studies* 2, no. 1 (1975): 09/03/29.

Jerven, Morten. *Poor Numbers: How We Are Misled by African Development Statistics and What to Do about It*. Ithaca: Cornell University Press, 2013.

Johnson, Franklyn A. 'The British Committee of Imperial Defence: Prototype of U.S. Security Organization'. *The Journal of Politics* 23, no. 2 (1961): 231–61.

Jones, G.S. *Outcast London: A Study in the Relationship between the Classes in Victorian Society*. New York: Pantheon, 1971.

Kahn, E. 'The Pass Laws'. In *Handbook of Race Relations in South Africa*, 275–91. Johannesburg: Race Relations Institute, 1949.

Katz, Elaine N. 'Revisiting the Origins of the Industrial Colour Bar in the Witwatersrand Gold Mining Industry, 1891–1899'. *Journal of Southern African Studies* 25, no. 1 (1999): 73–97.

Katz, Louis. 'Biometric Measuring Device'. New York, 15 May 1979.

Kaye, David H. 'Questioning a Courtroom Proof of the Uniqueness of Fingerprints'. *International Statistical Review* 71, no. 3 (1 December 2003): 521–33.

Kershaw, Sarah. 'Spain and US at Odds on Mistaken Terror Arrest'. *New York Times*, 5 June 2004.

Kevles, D.J. *In the Name of Eugenics: Genetics and the Uses of Human Heredity*. Cambridge, MA: Harvard University Press, 1995.

Kidd, Dudley. *Kafir Socialism and the Dawn of Individualism: An Introduction to the Study of the Native Problem*. London: A. and C. Black, 1908.

Killerlane III, J.J. 'Finger Imaging: A 21st Century Solution to Welfare Fraud at Our Fingertips'. *Fordham Urb. LJ Urban Law Journal* 22 (1994): 1327.

Kline, Benjamin. *Genesis of Apartheid: British African Policy in the Colony of Natal, 1845–1893*. Lanham, MD: University Press of America, 1988.

Knox, Robert. 'Inquiry into the Origin and Characteristic Differences of the Native Races Inhabiting the Extra-Tropical Part of Southern Africa'. *Memoirs of the Wernerian Natural History Society* V, no. 1 (1824).

  *The Races of Men: A Fragment*. London: H. Renshaw, 1850.

Koditschek, Theodore. *Liberalism, Imperialism, and the Historical Imagination: Nineteenth-Century Visions of a Greater Britain*. Cambridge University Press, 2011.

Koman, Richard. 'Server in the Sky: FBI International Biometric Db Planned'. *ZDNet*, 14 January 2008. http://government.zdnet.com/?p=3605.

*Kommissaris-Generaal: Xhosa Volkseenheid: Staatsdepartemente: Republiek Van Suid-Afrika. Justisie Landdroskantore Byhou Van Bevolkingsregister*. Vol. 28. KGT, 00000000.

Krasner, Jeffrey. 'S Africa on Cutting Edge of Bank Tech'. *Boston Herald*, 19 August 1995.

Kwankye, S.O. 'The Vital Registration System in Ghana: A Relegated Option for Demographic Data Collection'. In *The African Population in the 21st Century*, 420–43. Durban: Popline, 1999.

KwaZulu. *Report of the Committee of Enquiry into the Payment of Social Pensions in KwaZulu*, 1985.

Labour Correspondent. 'Survival Is Incidental to These "Shuffling Bureaucrats"'. *Rand Daily Mail*, 14 March 1980.

Lake, M. and H. Reynolds. *Drawing the Global Colour Line: White Men's Countries and the International Challenge of Racial Equality*. Cambridge University Press, 2008.

Lambert, John. *Betrayed Trust: Africans and the State in Colonial Natal*. Pietermaritzburg: University of Natal Press, 1995.

'From Independence to Rebellion: African Society in Crisis, 1880–1910'. In *Natal and Zululand from Earliest Times to 1910: A New History*, edited by Andrew Duminy and Bill Guest, 373–401. Pietermaritzburg: University of Natal, 1989.

'Last Rites for ID Cards Read by Johnson'. *Independent*, 1 July 2009. www. independent.co.uk/news/uk/home-news/last-rites-for-id-cards-readby-johnson-1726187.html.

'Latin America: Gini Back in the Bottle'. *The Economist*, 13 October 2012. www. economist.com/node/21564411.

Latour, Bruno. *Reassembling the Social: An Introduction to Actor-Network-Theory*. New York: Oxford University Press, 2005.

Lau, Brigitte. 'Conflict and Power in Nineteenth-Century Namibia'. *The Journal of African History* 27, no. 1 (1986): 29–39.

Lavin, D. *From Empire to International Commonwealth: A Biography of Lionel Curtis*. New York: Oxford University Press, 1995.

Layton, Edwin T. *The Revolt of the Engineers; Social Responsibility and the American Engineering Profession*. Cleveland: Press of Case Western Reserve University, 1971.

Lears, Jackson. *No Place of Grace: Antimodernism and the Transformation of American Culture, 1880–1920*. New York: Pantheon, 1981.

Leeman, Sue. 'Sad Saga of the Book of Life'. *Pretoria News*, 14 July 1983.

Legassick, Martin. 'British Hegemony and the Origins of Segregation in South Africa, 1901–1914'. In *Segregation and Apartheid in Twentieth Century South Africa*, edited by William Beinart and Saul Dubow, 43–59. London: Routledge, 1995.

Lehohla, Pali. *South African Statistics 2009*. Annual Report. South African Statistics. Pretoria: Statistics South Africa, 2009.

Lepore, Jill. *The Name of War: King Philip's War and the Origins of American Identity*. New York: Knopf, 1998.

'Let the House Ponder; Fingerprints Are Here'. *Weekly Mail*, 22 May 1986.

Levine, Philippa. 'Anthropology, Colonialism, and Eugenics'. In *The Oxford Handbook of the History of Eugenics*, 43–61. Oxford University Press, 2010.

Levy, S. *Progress against Poverty: Sustaining Mexico's Progresa-Oportunidades Program*. Washington, DC: Brookings Institution Press, 2006.

'Licences Warning to Blacks'. *Eastern Province Herald*, 25 January 1978.

Lichtenstein, A. 'Good Roads and Chain Gangs in the Progressive South: "The Negro Convict Is a Slave"'. *The Journal of Southern History* 59, no. 1 (1993): 85–110.

Lindert, K., A. Linder, J. Hobbs and B. de la Brière. 'The Nuts and Bolts of Brazil's Bolsa Família Program: Implementing Conditional Cash Transfers in a Decentralized Context', Washington, DC: The World Bank, 2006.

Lodge, Tom. *Black Politics in South Africa since 1945*. Johannesburg: Ravan Press, 1983.

*Sharpeville: An Apartheid Massacre and Its Consequences*. Oxford and New York: Oxford University Press, 2011.

Loveman, Mara. 'Blinded Like a State: The Revolt Against Civil Registration in Nineteenth-Century Brazil'. *Comparative Studies in Society and History* 49, no. 1 (2007): 5–39.

Lugg, Harry. Chief Native Commissioner, Natal. Letter to Secretary for Native Affairs, Pretoria. 'Registration of Births and Deaths amongst Natives', 1 July 1933. SAB NTS 9363 4/382 Natal and Zululand: Registration of births and deaths, 1911–1960.

Native Commissioner, Verulam. Letter to Secretary for Native Affairs, Pretoria. 'Registration of Births and Deaths amongst Natives', 7 June 1932. SAB NTS 9363 4/382 Natal and Zululand: Registration of births and deaths, 1911–1960.

Lund, Francie. *Changing Social Policy: The Child Support Grant in South Africa.* Pretoria: HSRC Press, 2007.

'Children, Citizenship and Child Support: The Child Support Grant in Post-Apartheid South Africa'. In *Registration and Recognition: Documenting the Person in World History*, edited by Keith Breckenridge and Simon Szreter, 495–514. Proceedings of the British Academy 182. Oxford University Press, 2012.

Lyon, D. *The Electronic Eye: The Rise of Surveillance Society.* Minneapolis: University of Minnesota Press, 1994.

*Surveillance after September 11.* Malden, MA: Polity Press in association with Blackwell Pub. Inc., 2003.

MacAskill, Ewen, Julian Borger, Nick Hopkins, Nick Davies and James Ball. 'GCHQ Taps Fibre-Optic Cables for Secret Access to World's Communications'. *Guardian*, 21 June 2013. www.guardian.co.uk/uk/2013/jun/21/gchq-cables-secret-world-communications-nsa.

Macgregor, David. 'Penis Print Scam Bust'. *Sunday Independent*, 8 June 1997.

Machiavelli, Niccolò. *The Prince.* Vol. 36. London: Oxford University Press, 1903.

MacKenzie, D.A. *An Engine, Not a Camera: How Financial Models Shape Markets.* Cambridge, MA: The MIT Press, 2006.

*Statistics in Britain, 1865–1930: The Social Construction of Scientific Knowledge.* Edinburgh: Edinburgh University Press, 1981.

MacKinnon, Aran S. 'The Persistence of the Cattle Economy in Zululand, South Africa, 1900–50'. *Canadian Journal of African Studies / Revue Canadienne Des Études Africaines* 33, no. 1 (1 January 1999): 98–135.

Macmillan, William M. *The Cape Colour Question: A Historical Survey.* 1st edn. New York: Humanities Press, 1927, reprinted 1968.

Magistrate, Kranskop. '1/KRK 3/1/4, KK931A/1902 Government Notice No. 799, 1902: Rules Re Registration of Births and Deaths of Natives', 1902. NAB.

Magnet, Shoshana. 'Bio-Benefits: Technologies of Criminalization, Biometrics and the Welfare System'. In *Surveillance and Social Problems*, 2008. www.magnetopia.org/biometrics%20and%20welfare.doc.

Maguire, Mark. 'The Birth of Biometric Security'. *Anthropology Today* 25, no. 2 (2009): 9–14.

Makinana, Andisiwe. 'Social Grants: 15.3-Million People to Register Again – News – Mail & Guardian Online'. *Mail and Guardian*, 15 February 2012. http://mg.co.za/article/2012-02-15-social-grants-153million-people-to-registeragain.

Mamdani, Mahmood. *Citizen and Subject: Contemporary Africa and the Legacy of Late Colonialism.* Cape Town: James Currey, 1996.

Mann, Michael. *The Sources of Social Power: The Rise of Classes and Nation-States, 1760–1914*. Vol. 2. Cambridge University Press, 1993.

Marable, Manning. *Malcolm X: A Life of Reinvention*. New York: Viking, 2011.

Marks, Shula. *Reluctant Rebellion: The 1906–8 Disturbances in Natal*. Oxford: Clarendon Press, 1970.

'South Africa's Early Experiment in Social Medicine: Its Pioneers and Politics'. *American Journal of Public Health* 87, no. 3 (1 March 1997): 452–9.

'War and Union, 1899–1910'. In *Cambridge History of South Africa, 1885–1994*, vol. 2, 157–210. Cambridge University Press, 2012.

Masote, Mamello. 'Absa Won't Quit Net1 Welfare Tender Battle'. *Business Day Live*, 31 March 2013. www.bdlive.co.za/business/financial/2013/03/31/absa-won-t-quit-net1-welfare-tender-battle.

Massey, D.S. and N.A. Denton. *American Apartheid*. Cambridge, MA: Harvard University Press, 1994.

Mavhunga, Clapperton. 'Firearms Diffusion, Exotic and Indigenous Knowledge Systems in the Lowveld Frontier, South Eastern Zimbabwe 1870–1920'. *Comparative Technology Transfer and Society* 1, no. 2 (2003): 201–31.

Mavrogordato, T.E. A/Chief Detective Inspector, CID. Letter to Commissioner of Police. 'Letter to Commissioner of Police', 19 June 1907. Tad Ld TAD LD 1466 Ag2497/07 Finger Impressions, 1907. TAD.

Mayer, Arno J. *Persistence of the Old Regime: Europe to the Great War*. New York: Pantheon Books, 1981.

Mazower, Mark. *No Enchanted Palace: The End of Empire and the Ideological Origins of the United Nations*. Lawrence Stone Lectures. Princeton University Press, 2009.

McClendon, Thomas V. 'Tradition and Domestic Struggle in the Courtroom: Customary Law and the Control of Women in Segregation-Era Natal'. *The International Journal of African Historical Studies* 28, no. 3 (1995): 527–61.

McClintock, Anne. *Imperial Leather*. London: Routledge, 1995.

McGerr, M.E. *A Fierce Discontent: The Rise and Fall of the Progressive Movement in America, 1870–1920*. New York: Free Press, 2003.

McKeown, Adam. *Melancholy Order: Asian Migration and the Globalization of Borders*. New York: Columbia University Press, 2008.

Meacham, Standish. *Toynbee Hall and Social Reform, 1880–1914: The Search for Community*. New Haven: Yale University Press, 1987.

Mears, Secretary for Native Affairs. Letter to Secretary for the Interior, Pretoria. 'Registration of Births and Deaths in Rural Native Areas', 10 May 1946. SAB NTS 9360 2/382 Part 2 Registration of Births and Deaths (natives) in the Union. (general File). 1947 to 1952.

Mehta, Uday Singh. *Liberalism and Empire: A Study in Nineteenth-Century British Liberal Thought*. University of Chicago Press, 1999.

Meissner, Doris and James Ziglar. 'The Winning Card'. *New York Times*, 16 April 2007. www.nytimes.com/2007/04/16/opinion/16meissner.html?scp=103&sq=biometric&st=cse.

Metcalf, Thomas R. *Imperial Connections: India in the Indian Ocean Arena, 1860–1920*. 1st edn. Berkeley: University of California Press, 2008.

Mignolo, Walter D. *Darker Side of the Renaissance: Literacy, Territoriality and Colonization*. Ann Arbor: University of Michigan, 1995.

Milner, Alfred. *England in Egypt*. London: E. Arnold, 1892.

Milner, High Commissioner, Cape Town. Letter to Lord Roberts, Pretoria. 'Telegram', 12 August 1900. TAD PSY 1 HC102/00 High Commissioner: Johannesburg Police Henry Appointed Temporary 1900–1900.

Letter to Lord Roberts, Pretoria. 'Telegram', 8 October 1900. TAD LD 15 LAM 1150/01 Commissioner of Police Reports Transvaal Police 1901.

Letter to Chamberlain, Secretary of State, London. 'Telegram', 20 December 1900. TAD LD 15 LAM 1150/01 Commissioner of Police Reports Transvaal Police 1901.

Mitchell, Timothy. *Carbon Democracy: Political Power in the Age of Oil*. London: Verso, 2011.

'The Limits of the State: Beyond Statist Approaches and Their Critics'. *The American Political Science Review* (1991): 77–96.

*Rule of Experts: Egypt, Techno-Politics, Modernity*. Berkeley: University of California Press, 2002.

'Society, Economy and the State Effect'. *The Anthropology of the State: A Reader* 9 (2006): 169.

'Society, Economy, and the State Effect'. In *State / Culture: State-Formation after the Cultural Turn*, edited by George Steinmetz, 76–97. Ithaca: Cornell University Press, 1999.

Mochiko, Thabiso. 'Net1 Given Leave to Appeal Social Grants Tender Verdict'. *Business Day Live*, 14 September 2012. www.bdlive.co.za/business/technology/2012/09/14/net1-given-leave-to-appeal-social-grants-tender-verdict.

Mongia, Radhika. 'Gender and the Historiography of Gandhian Satyagraha in South Africa'. *Gender & History* 18, no. 1 (2006): 130–49.

Moodie, T. Dunbar. 'The South African State and Industrial Conflict in the 1940s'. *The International Journal of African Historical Studies* 21, no. 1 (1 January 1988): 21–61.

Murray, Harry. 'Deniable Degradation: The Finger-Imaging of Welfare Recipients'. *Sociological Forum* 15, no. 1 (1 March 2000): 39–63.

Murray, John. 'Heunis Keeps a Finger on the Workforce'. *The Star*, 5 August 1980.

Musselman, Elizabeth Green. 'Swords into Ploughshares: John Herschel's Progressive View of Astronomical and Imperial Governance'. *The British Journal for the History of Science* 31, no. 4 (1 December 1998): 419–35.

'NAD MJPW 104, LW5435/1903 Agent General, London, Forwards a Letter from the Right Honourable Sir H Macdonell with Reference to the Prospects of Mr. Pinto Leite Obtaining a Commission in the Natal Police', 1903.

Nair, Janaki. 'Representing Labour in Old Mysore: Kolar Gold Fields Strike of 1930.' *Economic and Political Weekly* 25, no. 30 (July 28, 1990), 73–86.

Nandy, Ashis. 'From Outside the Imperium: Gandhi's Cultural Critique of the "West"'. *Alternatives* 7 (1981): 171–94.

'History's Forgotten Doubles'. *History and Theory* 34, no. 2 (May 1995): 44–66.

NARMIC. *Automating Apartheid: US Computer Exports to South Africa and the Arms Embargo*. Philadelphia: NARMIC, 1982.

Natal. Native Affairs Commission. *Evidence [of The] Native Affairs Commission, 1906–7*. Pietermaritzburg: P. Davis, 1907.

National Institute of Standards and Technology. 'Minutiae Interoperability Exchange Test 2004', 21 March 2006. http://fingerprint.nist.gov/minex04/.

Newmann, William W. 'Reorganizing for National Security and Homeland Security'. *Public Administration Review* 62, no. s1 (2002): 126–37.

Nilekani, N. *Imagining India: Ideas for the New Century*. London: Allen Lane, 2008.

Nimocks, Walter. *Milner's Young Men: The 'Kindergarten' in Edwardian Imperial Affairs*. London: Hodder & Stoughton, 1970.

Noble, David F. *America By Design: Science, Technology and the Rise of Corporate Capitalism*. Oxford University Press, 1977.

Norris, Christopher. *Derrida*. London: Fontana, 1987.

'Now Stand by for South Africa's Great Fingerprint Fiasco'. *Sunday Express*, 18 January 1981.

Ntsebeza, Lungisile. *Democracy Compromised: Chiefs and the Politics of Land in South Africa*. Cape Town: HSRC Press, 2006.

O'Meara, Dan. *Forty Lost Years: The Apartheid State and the Politics of the National Party, 1948–1994*. Johannesburg: Ravan Press, 1996.

Overy, R.J. *The Morbid Age: Britain between the Wars*. London: Allen Lane, 2009.

Page, Lewis. 'UK.gov Says No Plans for FBI DNA Database Hookup'. *The Register*, 17 January 2008. www.theregister.co.uk/2008/01/17/fbi_uk_dna_database_plans_followup/.

Palmer, Rebecca. 'NZ Police May Join FBI Network'. *Stuff.co.nz*, 15 September 2008. www.stuff.co.nz/4357650a11.html.

Parker, Geoffrey. *The Grand Strategy of Philip II*. New Haven: Yale University Press, 1998.

Pearse, Henry. *The History of Lumsden's Horse; a Complete Record of the Corps from Its Formation to Its Disbandment*. London: Longmans, Green & Co, 1905. www.archive.org/details/historyoflumsden00pearrich.

Pearson, Karl. *The Life, Letters and Labours of Francis Galton: Birth 1822 to Marriage 1853*. 4 vols, vol. 1. Cambridge University Press, 1914.

*The Life, Letters and Labours of Francis Galton: Correlation, Personal Identification and Eugenics*. 4 vols, vol. 3A. Cambridge University Press, 1930.

*The Life, Letters and Labours of Francis Galton: Researches of Middle Life*. 4 vols, vol. 2. Cambridge University Press, 1924.

*National Life from the Standpoint of Science*. London: A & C Black, 1900. www.archive.org/details/nationallifefrom00pearrich.

Peires, Jeff B. *The Dead Will Arise: Nongqawuse and the Great Xhosa Cattle-Killing Movement of 1856–7*. Johannesburg: Ravan Press, 1989.

Penrose, Paul. 'Out of the Backwoods'. *The Times*, 10 February 1995.

'Pensioners Asked to Pay "Special Tax"'. *The New Nation*, 9 May 1991.

Peterson, Derek R. *Creative Writing: Translation, Bookkeeping, and the Work of Imagination in Colonial Kenya*. Portsmouth, NH: Heinemann, 2004.

Piazza, Pierre and Laurent Laniel. 'The INES Biometric Card and the Politics of National Identity Assignment in France'. In *Playing the Identity Card: Surveillance, Security and Identification in Global Perspective*, edited by David Lyon and Colin Bennett, 93–111. London and New York: Routledge, 2008.

Pinney, C. *Camera Indica: The Social Life of Indian Photographs*. University of Chicago Press, 1997.

Pinto-Leite, N.P. 'The Finger-Prints'. *The Nonqai* (1907): 31–3.

Polakow-Suransky, Sasha. *The Unspoken Alliance: Israel's Secret Relationship with Apartheid South Africa*. New York: Pantheon, 2010.

Political Staff. 'Why Black Finger Prints?' *Cape Times*, 11 March 1980.

Pollack, Andrew. 'Recognizing the Real You'. *New York Times*, 24 September 1981.

'Population Registration Amendment Bill. Second Reading'. *Hansard*, 17 March 1967.

Porter, Bernard. *The Origins of the Vigilant State: The London Metropolitan Police Special Branch before the First World War*. London: Boydell & Brewer Inc, 1991.

Porter, Theodore M. *Karl Pearson: The Scientific Life in a Statistical Age*. Princeton University Press, 2004.

*The Rise of Statistical Thinking 1820–1900*. Princeton University Press, 1986.

*Trust in Numbers: The Pursuit of Objectivity in Science and Public Life*. Princeton University Press, 1995.

Posel, Deborah. *The Making of Apartheid, 1948–1961: Conflict and Compromise*. Oxford and New York: Clarendon Press and Oxford University Press, 1991.

'Race as Common Sense: Racial Classification in Twentieth-Century South Africa'. *African Studies Review* 44, no. 2 (2001): 87–113.

'What's in a Name? Racial Categorisations under Apartheid and Their Afterlife'. *Transformation* (2001): 50–74.

Pottinger, Brian. 'The Fall of an Empire That Ruled Black Lives'. *Sunday Times*, 10 November 1985.

Power, Paul F. 'Gandhi in South Africa'. *The Journal of Modern African Studies* 7, no. 3 (1969): 441–55.

Powers, Richard Gid. *Secrecy and Power: The Life of J Edgar Hoover*. London: Hutchinson & Co, 1987.

'Presentation of the Gold Medals'. *Journal of the Royal Geographical Society of London* 23 (1853): lviii–lxi.

Progressus Research Development Consultancy. *The Payment Experience of Social Grant Beneficiaries*. FinMark Trust, April 2012.

Prozesky, M. and J.M. Brown. *Gandhi and South Africa: Principles and Politics*. Pietermaritzburg: University of Natal Press, 1996.

'R700000 for Armoured Cars'. *The Citizen*, 16 September 1993.

Radcliffe-Brown, Alfred. 'Preface'. In *African Political Systems*, edited by Edward Evan Evans-Pritchard, Meyer Fortes and Rachel Dempster, xi–xxiii. Oxford: International African Institute, 1955. http://library.wur.nl/WebQuery/clc/464172.

Rampedi, Piet. 'Battle for R10bn Tender'. *The Star*, 29 May 2012.

Rauchway, E. *Blessed Among Nations: How the World Made America*. New York: Hill and Wang, 2007.

'Reisdokument Gaan Pasboek Vervang'. *Die Volksblad*, 23 July 1977.

Republic of South Africa. *Bantu Homelands Citizenship Act*, 1970.

*Request of Belgian Government to Be Supplied with Information Regarding the Finger Print System*. Vol. 111. GG, 1912.

Rich, Paul B. and Richard Stubbs. *The Counter-Insurgent State: Guerrilla Warfare and State Building in the Twentieth Century*. London: Macmillan, 1997.

Richardson, Peter. *Chinese Mine Labour in the Transvaal*. London: Macmillan, 1982.

Rodgers, D.T. *Atlantic Crossings: Social Politics in a Progressive Age*. Cambridge, MA: Harvard University Press, 2000.

Rodney, Walter. *How Europe Underdeveloped Africa*. Washington: Howard University Press, 1982.

Rodriguez, Julia. 'South Atlantic Crossings: Fingerprints, Science, and the State in Turn-of-the-Century Argentina'. *The American Historical Review* 109, no. 2 (2004): 1–42.

Rosental, Paul-André. 'Civil Status and Identification in Nineteenth-Century France: A Matter of State Control?' In *Registration and Recognition: Documenting the Person in World History*, edited by Keith Breckenridge and Simon Szreter, 137–65. Proceedings of the British Academy 182. Oxford University Press, 2012.

Rossouw, Rehana. 'The Payment That Put Paid to Abe'. *Weekly Mail*, 23 February 1996.

Roy, Tirthankar. 'Indigo and Law in Colonial India'. *The Economic History Review* 64 (2011): 60–75.

Ruggiero, K. 'Fingerprinting and the Argentine Plan for Universal Identification in the Late Nineteenth and Early Twentieth Centuries'. In *Documenting Individual Identity: The Development of State Practices in the Modern World*, edited by Jane Caplan and John Torpey, 184–96. Princeton University Press, 2001.

Ruskin, John. *Unto This Last*. London: Cornhill Magazine, 1860.

Russell, Mark. 'FBI Invites Australia to Join World Crime Database'. *The Age*, 20 January 2008. www.theage.com.au/news/national/fbi-invites-australia-to-join-world-crime-database/2008/01/19/1200620280804.html.

'Rybewyse Is Die Boek Se Swak'. *Beeld*, 4 July 1978.

Rycroft, Alan. 'Social Pensions and Poor Relief: An Exercise in Social Control'. Unpublished, 1987.

Sagner, Andreas. 'Ageing and Social Policy in South Africa: Historical Perspectives with Particular Reference to the Eastern Cape'. *Journal of Southern African Studies* 26, no. 3 (September 2000): 523–53.

Sankar, P. 'State Power and Record-Keeping: The History of Individualized Surveillance in the United States, 1790–1935'. University of Pennsylvania, 1992.

SAPA. 'CPA Fingerprint Pension Move'. *The Citizen*, 16 November 1993.

'Sassa Judgment Illegal but Won't Be Set Aside'. *The M&G Online*, 28 August 2012. http://mg.co.za/article/2012-08-28-sassa-ruling-illegal-but-wont-be-set-aside/.

Sassen, Saskia. *The Global City: New York, London, Tokyo*. Princeton University Press, 1991.

Savage, Michael. 'The Imposition of Pass Laws on the African Population in South Africa 1916–1984'. *African Afairs* 85, no. 339 (1986): 181–205.

'SBSA – Banking the Unbanked'. *Computerworld Honors Program*, 1998. www. cwheroes.org/Search/his_4a_detail.asp?id=3453.

Schrecker, Ellen. *Many Are the Crimes: McCarthyism in America*. Boston: Little Brown and Company, 1998.

Scott, James C. *The Art of Not Being Governed: An Anarchist History of Upland Southeast Asia*. New Haven: Yale University Press, 2009.

   *Seeing like a State: How Certain Schemes to Improve the Human Condition Have Failed*. New Haven: Yale University Press, 1998.

Searle, G.R. *Eugenics and Politics in Britain, 1900–1914*. Leyden: Noordhoff International Publishing, 1976.

   *The Quest for National Efficiency: A Study in British Politics and Political Thought, 1899–1914*. Oxford: Basil Blackwell, 1971.

*Secondment of Fingerprint Experts to Kenya*. Vol. 512. SAP, 1954.

Secretary for Interior. Letter to Secretary for Native Affairs, Pretoria, 15 October 1913. SAB NTS 9363 3/382 Transkeian territories: Registration of births and deaths, 1911–1960.

Secretary for Interior, Pretoria. Letter to Secretary for Native Affairs, Pretoria, 10 March 1913. SAB NTS 9363 4/382 Natal and Zululand: Registration of births and deaths, 1911–1960.

Secretary for Native Affairs. *Circular: Registration of Birth*. Vol. 3/1/2. 1/ KRK, 1898.

   'SAB NTS 9360 2/382 Part 2 Registration of Births and Deaths (natives) in the Union. (general File). 1947 to 1952', 1952 1947. Central Archives, Pretoria. www.national.archsrch.gov.za/sm300cv/smws/sm30ddf0?2008063 02009567CA78806&DN=00000002.

Secretary for Native Affairs, Pretoria. 'Memorandum: Geddes Axe Committee: Registration of Native Births and Deaths in Natal', 31 October 1922. SAB NTS 9363 4/382 Natal and Zululand: Registration of births and deaths, 1911–1960.

Seegers, Annette. 'One State, Three Faces: Policing in South Africa (1910–1990)'. *Social Dynamics* 17, no. 1 (1991): 36–48.

Seekings, Jeremy. 'Visions, Hopes & Views about the Future: The Radical Moment of South African Welfare Reform'. In *South Africa's 1940s: Worlds of Possibilities*, edited by Saul Dubow and Alan Jeeves. jutaonline.co.za. Cape Town: Juta, 2005.

Sekula, Allan. 'The Body and the Archive'. *October* 39 (1986): 3–64.

Semmel, Bernard. *Imperialism and Social Reform: English Social-Imperial Thought, 1895–1914*. London: George Allen & Unwin, 1960.

   *Jamaican Blood and Victorian Conscience: The Governor Eyre Controversy*. New York: Houghton Mifflin, 1963.

Sengoopta, Chandak. *Imprint of the Raj: How Fingerprinting Was Born in Colonial India*. London: Macmillan, 2003.

Senior Assistant Health Officer. Letter to Secretary for Public Health, 13 March 1933. SAB NTS 9363 4/382 Natal and Zululand: Registration of births and deaths, 1911–1960.

Shapiro, S. 'Development of Birth Registration and Birth Statistics in the United States'. *Population Studies* 4, no. 1 (1950): 86–111.

Shepstone, Arthur (Actg. Under Secretary for Native Affairs, Province of Natal). Letter to Acting Secretary for Native Affairs, Pretoria. 'Registration of Births and Deaths (Natives in Zululand)', 29 May 1911. SAB NTS 9363 4/382 Natal and Zululand: Registration of births and deaths, 1911–1960.

Singha, Radhika. 'Settle, Mobilize, Verify: Identification Practices in Colonial India'. *Studies in History* 16, no. 2 (2000): 151–98.

Sinha, M. *Colonial Masculinity: The 'Manly Englishman' and the 'Effeminate Bengali' in the Late Nineteenth Century*. Manchester: Manchester University Press, 1995.

Skinner, Quentin. *Machiavelli: A Very Short Introduction*. Oxford University Press, 2000.

Skocpol, Theda. *Protecting Soldiers and Mothers: The Political Origins of Social Policy in the United States*. Cambridge, MA: Harvard University Press, 1992.

Smit, D.L., Secretary for Native Affairs, Cape Town. Letter to Harry Lugg, Chief Native Commissioner, Natal. 'Registration of Births and Deaths of Natives,' February 26, 1935. SAB NTS 9363 4/382 Natal and Zululand: Registration of births and deaths, 1911–1960.

Smith, Anna Marie. *Welfare Reform and Sexual Regulation*. Cambridge University Press, 2007.

Smuts, H.H.L. Letter to Secretary for Native Affairs. 'Memo', 10 March 1949. SAB NTS 9360 2/382 Part 2 Registration of Births and Deaths (natives) in the Union. (general File). 1947 to 1952.

Smuts, J.C. Letter to Merriman, J X, 8 January 1908. Selections from the Smuts Papers.

Solomon, Richard. 'Enclose in No 15. Minute. Penalties under the Pass Law and Gold Law of the Transvaal', 15 July 1901. Cd 904 Papers relating to legislation affecting Natives in the Transvaal (in continuation of [Cd 714], July 1901).

Letter to Fiddes, W, 2 February 1901. TAD LD 15 LAM 1150/01 Commissioner of Police Reports Transvaal Police 1901.

Solove, Daniel J. '"I've Got Nothing to Hide" and Other Misunderstandings of Privacy'. *San Diego Law Review* 44 (2007): 745–72.

South African Government. 'Green Paper on Families: Promoting Family Life and Strengthening Families in South Africa', 12 August 2012. www.thepresidency.gov.za/pebble.asp?relid=6654.

Sowetan Correspondent. 'Pensions Backlog: KwaZulu's Old Age Budget Has Dried Up'. *The Sowetan*, 23 January 1984.

Spence, Clark C. *Mining Engineers & the American West; the Lace-Boot Brigade, 1849–1933*. New Haven: Yale University Press, 1970.

Spencer, Herbert. *Principles of Sociology: Volume 1*. New York: D Appleton and Company, 1912.

*The Study of Sociology*. New York: D Appleton and Company, 1878.

Staff Reporter. '8-M Blacks Get Identity Docs'. *The Citizen*, 18 February 1991.

'Absa, AllPay Lose Court Bid over Tender'. *Sunday World*, 1 April 2013. www.sundayworld.co.za/news/2013/04/01/absa-allpay-lose-court-bid-over-tender.

'Rush for New Non-Racial ID Documents'. *The Star*, 7 January 1987.

'Tokyo Sexwale's Prints All over R10-Billion Tender'. *The M&G Online*, 23 March 2012. http://mg.co.za/article/2012-03-23-sexwales-prints-all-over-r10bn-tender/.

Staff Writer. 'CPS in Loan Probe'. *ITWeb Business*, 27 May 2013. www.itweb. co.za/index.php?option=com_content&view=article&id=64399:CPS-in-loan-probe&catid=69.

Stanford, S.H. Chief Magistrate, Transkeian Territories. Letter to Dower, Secretary for Native Affairs, 3 March 1911. SAB NTS 9363 3/382 Transkeian territories: Registration of births and deaths, 1911–1960.

Steinmetz, G. *The Devil's Handwriting: Precoloniality and the German Colonial State in Qingdao, Samoa, and Southwest Africa*. University of Chicago Press, 2007.

Stepan, Nancy. *The Idea of Race in Science: Great Britain, 1800–1960*. Basingstoke: Macmillan in association with St Antony's College Oxford, 1982.

Stigler, S.M. *The History of Statistics: The Measurement of Uncertainty before 1900*. New York: Belknap Press, 1986.

Stirling, Tony. 'Black Sash Claims on Black IDs Repudiated'. *The Citizen*, 8 January 1988.

Stocking, George W. *Victorian Anthropology*. New York: Free Press, 1991.

Stoler, Ann Laura. *Along the Archival Grain: Epistemic Anxieties and Colonial Common Sense*. Princeton University Press, 2008.

Stubbs, R. 'The Malayan Emergency and the Development of the Malaysian State'. *The Counter-Insurgent State: Guerrilla Warfare and State Building in the Twentieth Century* (1997): 50–71.

Stultz, Newell M. 'Evolution of the United Nations Anti-Apartheid Regime'. *Human Rights Quarterly* 13, no. 1 (1 February 1991): 1–23.

Subcommittee on Biometrics. *The National Biometrics Challenge*. Washington, DC: National Science and Technology Council, 15 September 2011. www. biometrics.gov/NSTC/Publications.aspx.

Swan, M. *Gandhi: The South African Experience*. Johannesburg: Ravan Press, 1985.

Swanson, Maynard W. '"The Asiatic Menace": Creating Segregation in Durban, 1870–1900'. *The International Journal of African Historical Studies* 16, no. 3 (1983): 401–21.

Swilling, Mark and Mark Phillips. 'State Power in the 1980s: From "Total Strategy" to "Counter-Revolutionary Warfare"'. In *War and Society*, edited by Jacklyn Cock and Laurie Nathan, 134–48. Cape Town: David Philip, 1989.

Szreter, Simon. *Fertility, Class and Gender in Britain, 1860–1940*. Cambridge University Press, 2002.

'Registration of Identities in Early Modern English Parishes and Amongst the English Overseas'. In *Registration and Recognition: Documenting the Person in World History*, edited by Keith Breckenridge and Simon Szreter, 67–92. Proceedings of the British Academy 182. Oxford University Press, 2012.

Szreter, Simon and Keith Breckenridge. 'Recognition and Registration: The Infrastructure of Personhood in World History'. In *Registration and Recognition: Documenting the Person in World History*, edited by Keith Breckenridge and Simon Szreter, 1–36. Proceedings of the British Academy 182. Oxford University Press, 2012.

'TAB SP 177 Spr6780/98 Kamer Van Mynwezen Johannesburg. Re Detentie Van Naturellen Voor Identificatie. 18980715 to 18980715', n.d.

'TAB SS 0 R13775/98 Publieke Aanklager Pretoria. Geeft Wyziging Aan Den Hand Van Wet 2/1898 (paswet). 18981019 to 18981019 See R5353/98', n.d. www.national.archsrch.gov.za/sm300cv/smws/sm30ddf0?20080629165354 5A3C588C&DN=00000082.

Taussig, Michael. *Mimesis and Alterity: A Particular History of the Senses*. New York: Routledge, 1993.

Tennant. Secretary to Law Department. Letter to Assistant Colonial Secretary, Pretoria. 'Letter to Assistant Colonial Secretary', 19 June 1907. LD 1466 AG2497/07 Finger Impressions, 1907. TAD.

Theron, Jaap. 'Influx Change Expected Soon'. *The Citizen*, 3 November 1977.

Tignor, Robert L. 'The "Indianization" of the Egyptian Administration under British Rule'. *American Historical Review* 68, no. 3 (1963): 636–61.

Tilley, Helen. *Africa as a Living Laboratory: Empire, Development, and the Problem of Scientific Knowledge, 1870–1950*. University of Chicago Press, 2011.

Tilly, Charles. *Coercion, Capital and European States, AD 990–1992*. Oxford: Blackwell, 1992.

Tonkin, Arnold H., Hon Sec, South African Medical Association, Border Branch, Transkei Division. Letter to H Lawrence, Minister of Interior. 'Registration of Births and Deaths in Rural Native Areas', 28 April 1942. SAB NTS 9360 2/382 Part 2 Registration of Births and Deaths (natives) in the Union. (General file). 1947 to 1952.

Torpey, John. *The Invention of the Passport: Surveillance, Citizenship and the State*. Cambridge University Press, 2000.

*Transkei. Binnelandse Sake. Bevolkingsregistrasie*. Vol. 12/194. BAO, 1968.

*Transmits Request of Belgian Govt. to Be Supplied with Certain Information as to the Working of the Fingerprint System in South Africa*. Vol. 109. GG, 1912.

Truter, Theo G. Acting Chief Commissioner of the South African Police. Letter to Acting Secretary for Justice. 'Finger Print System in South Africa', 27 May 1912. CAD JUS 0862, 1/138. URL.

Turner, Frank M. 'Victorian Scientific Naturalism and Thomas Carlyle'. *Victorian Studies* 18, no. 3 (March 1975): 325–43.

'United States Visa Regulations'. *Hansard*, UK, 23 March 1954. http://hansard.millbanksystems.com/lords/1954/mar/23/united-states-visa-regulations#S5LV0186P0_19540323_HOL_13.

'US Diplomats Spied on UN Leadership'. *Guardian*, 28 November 2010. www.guardian.co.uk/world/2010/nov/28/us-embassy-cables-spying-un.

Vaderlandspan. 'Bewysboek Verwelkom'. *Die Vaderland*, 15 January 1981.

Van der Horst, Sheila T. *Native Labour in South Africa*. Oxford University Press, 1942.

Van Onselen, Charles. 'Crime and Total Institutions in the Making of Modern South Africa: The Life of "Nongoloza" Mathebula, 1867–1948'. *History Workshop Journal* 19 (1985): 62–81.

*The Fox and the Flies: The World of Joseph Silver, Racketeer and Psychopath*. London: Jonathan Cape, 2007.

*Studies in the Social and Economic History of the Witwatersrand, 1886–1914: Volume 1 New Babylon.* Johannesburg: Ravan Press, 1982.

'Who Killed Meyer Hasenfus? Organized Crime, Policing and Informing on the Witwatersrand, 1902–8'. *History Workshop Journal* 67, no. 1 (20 March 2009): 1–22.

van Schalkwyk, Minister, South Africa House. Letter to Secretary for External Affairs. 'Parliamentary Question: Fingerprinting of British Subjects Applying for Visas', 4 December 1957. BNS 1/1/328, 42/74 Fingerprints and Photographs on Permits Etcetera Issued to Indians. General Questions. Part 2. 1928–1957. SAB.

Vansina, Jan. *Paths in the Rainforest: Toward a History of Political Tradition in Equatorial Africa.* Madison: University of Wisconsin Press, 1990.

Vaughan, Megan. *Curing Their Ills: Colonial Power and African Illness.* Stanford: Stanford University Press, 1991.

*Venda. Bevolkingsregistrasie. Sensus En Statistiek.* Vol. 11/76. BAO, 1979.

'Vingerafdrukke'. *Die Afrikaner*, 23 January 1981.

'Vingerafdrukke Gevra'. *Die Burger*, 15 January 1981.

Viroli, Maurizio. *From Politics to Reason of State: The Acquisition and Transformation of the Language of Politics 1250–1600.* Cambridge University Press, 2005.

'Visa Formalities (Finger Prints)'. *Hansard*, UK, 8 March 1954. http://hansard.millbanksystems.com/commons/1954/mar/08/visa-formalities-finger-prints#S5CV0524P0_19540308_HOC_94.

*Visit of JD Grant from Kenya Colony for Instruction in Finger Prints at Criminal Record Bureau Pretoria.* Vol. 1/2/278. PM, 1930.

Visser, Amanda. '"No Irregularity" in R10bn Cash Paymaster Tender Win'. *Business Day Live*, 28 March 2013. www.bdlive.co.za/national/2013/03/28/no-irregularity-in-r10bn-cash-paymaster-tender-win.

Von Glahn, Richard. 'Household Registration, Property Rights, and Social Obligations in Imperial China: Principles and Practices'. In *Registration and Recognition: Documenting the Person in World History*, edited by Keith Breckenridge and Simon Szreter, 39–66. Proceedings of the British Academy 182. Oxford University Press, 2012.

Walker, C. *Women and Resistance in South Africa.* New York: Monthly Review Press, 1991.

Warner, Michael. *The Letters of the Republic: Publication and the Public Sphere in Eighteenth Century America.* Cambridge, MA: Harvard University Press, 1990.

Warwick, Peter. *Black People and the South African War (1899–1902).* Cambridge University Press, 1980.

Webb, Beatrice Potter. *All the Good Things of Life, 1892–1905*, edited by Norman Ian MacKenzie and Jeanne MacKenzie. Cambridge, MA: Harvard University Press, 1983.

*'Glitter Around and Darkness Within,' 1873–1892*, edited by Jeanne MacKenzie and Norman Ian MacKenzie. Cambridge, MA: Belknap Press of Harvard University Press, 1982.

*'The Power to Alter Things,' 1905–1924*, edited by Norman Ian MacKenzie and Jeanne MacKenzie. Cambridge, MA: Belknap Press of Harvard University Press, 1984.

Weber, Max. *Economy and Society*, edited by Guenther Roth and Claus Wittich. 2 vols. Berkeley: University of California Press, 1978.

Wells, Julia C. *We Now Demand! The History of Women's Resistance to Pass Laws in South Africa*. Johannesburg: Witwatersrand University Press, 1993.

Welsh, David John. *The Roots of Segregation: Native Policy in Natal (1845–1910)*. Cape Town: Oxford University Press, 1971.

Whither Biometrics Committee, National Research Council. *Biometric Recognition: Challenges and Opportunities*, edited by Joseph N. Pato and Lynette I. Millett. Washington, DC: The National Academies Press, 2010.

'Why Fuss about Fingerprinting?' *The Natal Mercury*, 26 April 1986.

Wiebe, Robert H. *The Search for Order, 1877–1920*. New York: Hill and Wang, 1967.

Wiener, Norbert. *The Human Use of Human Beings: Cybernetics and Society*. Cambridge, MA: Da Capo Press, 1988.

'WikiLeaks Raises Specter Of Biometric Data'. *All Things Considered*. NPR, 30 November 2010. www.npr.org/2010/11/30/131704360/wikileaks-raises-specter-of-biometric-data.

Willcocks, W. *Mr Willcocks' Report on Irrigation in South Africa*. BPP Cd 1163 Further Correspondence Relating to Affairs in South Africa. (In Continuation of [Cd 903] January 1902). Daira Sania Co, Egypt, 1901.

Wilmsen, Edwin N. *Land Filled with Flies: A Political Economy of the Kalahari*. University of Chicago Press, 1989.

Wilson, Dean. 'The Politics of Australia's "Access Card"'. In *Playing the Identity Card: Surveillance, Security and Identification in Global Perspective*, edited by David Lyon and Colin Bennett, 180–97. London and New York: Routledge, 2008.

Wilson, Edward, Superintendent FIRD. Letter to H S Cooke, Acting Pass Commissioner. 'Report Working of the NAD Pass Office FIRD', 12 February 1907. TAD SNA 354 The Acting Pass Commissioner, Johannesburg, forwards copy of a report by the superintendent, Finger Impression Record Department, 1907.

Superintendent FIRD, Pass Office, Johannesburg. Letter to District Controller, May 2, 1906. TAD SNA 332 Na2491/06 Part 1 Proposed Extension of Finger Impression Record Department Na2782/06 Part 1 the Under Secretary Division 2 Colonial Secretary's Office Finger Print Impression Experts 1906.

Wilson, Francis. 'Unresolved Issues in the South African Economy: Labour'. *South African Journal of Economics* 43, no. 4 (1975): 311–27.

Wilson, Joan Hoff. *Herbert Hoover, Forgotten Progressive*. Boston: Little, Brown, 1975.

Wilson, Susan R. 'Evolution and Biometry'. *Biometrics* 55, no. 2 (June 1999): 333–7.

Witwatersrand Chamber of Mines. *The Mining Industry, Evidence and Report of the Industrial Commission of Enquiry, with an Appendix*. Johannesburg: Chamber of Mines, 1897.

Wood, David Murakami and Rodrigo Firmino. 'Empowerment or Repression? Opening up Questions of Identification and Surveillance in Brazil through a Case of "Identity Fraud"'. *Identity in the Information Society* 2, no. 3 (1 December 2009): 297–317.

Woodside, Alexander. *Lost Modernities: China, Vietnam, Korea, and the Hazards of World History*. Cambridge, MA: Harvard University Press, 2006.

Woodward, C. Vann. *The Strange Career of Jim Crow*. New York: Oxford University Press, 1966.

Woollacott, Angela. *Gender and Empire*. London: Palgrave Macmillan, 2006.

Wright, Marcia. 'Public Health among the Lineaments of the Colonial State in Natal, 1901–1910'. *Journal of Natal and Zulu History* 24 (2007): 135–61.

Young, M. Crawford. *The African Colonial State in Comparative Perspective*. New Haven: Yale University Press, 1994.

'Zac De Beer: Dr Zackyll and Mr (Take You For A) Ride?' *Noseweek*, July 1993.

Zelazny, F. 'The Evolution of India's UID Program' (2012). http://cgdev.org/files/1426371_file_Zelazny_India_Case_Study_FINAL.pdf.

Zimmermann, Eduardo A. 'Racial Ideas and Social Reform: Argentina, 1890–1916'. *The Hispanic American Historical Review* 72, no. 1 (1992): 23–46.

'Zulus Appoint Probe into Pension Schemes'. *The Daily News*, 11 April 1984.

# Index

Printed in the United States
By Bookmasters